Mechanical Engineering Technologies and Applications

(Volume 1)

Edited by

Zied Driss
Laboratory of Electromechanical Systems (LASEM)
National School of Engineers of Sfax (ENIS)
University of Sfax (US)
Sfax
Tunisia

Mechanical Engineering Technologies and Applications

Editor: Zied Driss

ISBN (Online): 978-981-4998-18-5

ISBN (Print): 978-981-4998-19-2

ISBN (Paperback): 978-981-4998-20-8

©2021, Bentham Books imprint.

Published by Bentham Science Publishers Pte. Ltd. Singapore. All Rights Reserved.

need for a court order if at any point you breach any terms of this License Agreement. In no event will any delay or failure by Bentham Science Publishers in enforcing your compliance with this License Agreement constitute a waiver of any of its rights.

3. You acknowledge that you have read this License Agreement, and agree to be bound by its terms and conditions. To the extent that any other terms and conditions presented on any website of Bentham Science Publishers conflict with, or are inconsistent with, the terms and conditions set out in this License Agreement, you acknowledge that the terms and conditions set out in this License Agreement shall prevail.

Bentham Science Publishers Pte. Ltd.
80 Robinson Road #02-00
Singapore 068898
Singapore
Email: subscriptions@benthamscience.net

BENTHAM SCIENCE

CONTENTS

PREFACE

This book focuses on the dissemination of information of permanent interest in mechanic applications and engineering technology. The considered applications are widely used in several industrial fields, particularly in those of automotive and aerospace aspects. Many features related to Mechanic processes are presented. The presented case studies and development approaches aim to provide the readers, such as engineers and PhD students, with basic and applied studies broadly related to Mechanic Applications and Engineering Technology.

In the first chapter, thetwisted Darrieus turbine is suggested as an amelioration of the conventional Darrieus rotor by modifying it to have helical blades. This reform affords the twisted turbine's better performances with regard to the conventional turbine. For this, a computational study of a twisted Darrieus rotor is conducted through the unsteady Reynolds-Averaged Navier-Stokes (URANS) equations. Different grid sizes are investigated to assess the impact of grid generation on the computing findings. The validation of the computing method with antecedent tests is carried out to select the adequate grid size. The flow characteristics of the water around the twisted Darrieus rotor have been assessed and discussed.

In the second chapter, the purpose is to study the heat exchange which is directly related to factors such as Reynolds number, thermal properties of materials, geometric shapes and dimensions. A numerical study of the heat exchanges between cross-sections selected of a mini-channel cooler is carried out. Three different forms have been considered for cooling an electronic component using a nanofluid (CuO-water) as a cooling liquid with a 4% volume concentration of nanoparticles. The simulation is carried out using the ANSYS Fluent software. The Reynolds number (Re) is taken between 100 and 700, and the stream regime is assumed to be stationary. The results obtained for the three forms of mini-channels proposed show that the rise in the exchange surface between the CuO-water nanofluid and walls of the mini-channels leads to the increase in the heat exchange coefficient and to the amelioration of the maximum temperature of electronic components by increasing the value of the flow velocity. This is confirmed by the results of the third case. In contrast to the first case that does not contain ribs, and the second case, which contains two ribs inside the channel, these two cases provide insufficient heat exchange, and the maximum temperature of the electronic component remains high compared to the third case, which contains four ribs, the latter contributes to the increase in heat exchange inside the channel.

The complexity of orthodontic treatments requires archwires with specific biomechanical properties according to the different stages of therapy. Thanks to their wide elastic zone and low stiffness, superelastic NiTi alloy is used in the leveling and alignment phases. The friction that accompanies the beginning of treatment is a very complicated phenomenon, since in the presence of arch misalignments, the present normal force, which compresses the orthodontic archwire-bracket couple, is very dependent on the clinical situation. The third chapter aims to identify the friction responses and the degradation mechanisms of a superelastic NiTi orthodontic archwire as a function of the applied normal load. The latter represents the charges delivered by the archwire during its unloading, all through the first phases of treatment. Circular and rectangular samples with the most commonly used dimensions have been tested in a dry environment at room temperature. The results confirm that the wear of the NiTi alloy is amplified as a function of the normal force applied for the two tested archwire shapes. Indeed, the degradation regimes observed by scanning electron microscopy present a transition by increasing the load from a mainly adhesive regime to a more complex situation, in which wear by adhesion is accompanied by abrasive and delamination wear.

In the fourth chapter, the influence of the Reynolds number on the flow around a Savonius wind rotor is investigated. Particularly, various regimes defined by different Reynolds number values equal to Re = 98000, Re = 111000, Re = 124000 and Re = 137000 are considered. For this, an open wind tunnel is used to evaluate the global characteristics of the Savonius wind rotor. The overall performance evaluation of the rotor is focused on the power, the dynamic and the static torque coefficient evolution.

In the fifth chapter, Zn-Mn electrodeposition from additive-free chloride bath on steel is investigated. Several operating parameters, namely the Mn2 concentration, the current density and the stirring, are explored with regard to the Mn content in the Zn matrix. The Mn content depends on the applied current density and jumps from zero to a maximum of 11.4% under 140 mA/cm^2. At high current density, Zn-Mn coatings are darker, more dendritic and with bad adhesion to the substrate. The dark appearance of Zn-Mn alloys is linked to oxy/hydroxide inclusions formed into the co-deposits.

In the sixth chapter, divalent europium-activated alpha-distrontium diphosphate (α-S2P2O7) phosphor powders are successfully prepared by a conventional solid-state reaction method under a reduced atmosphere. Synthesized samples are characterized by means of X-ray diffraction (XRD) patterns, nuclear magnetic resonance (NMR) and infrared (IR) spectroscopy which signify the formation of a pure single phase of Sr2P2O7. The optical properties are studied in both the ultraviolet (UV) and vacuum ultraviolet (VUV) regions. The emission spectra are obtained by excitation at 131 or 320 nm, whichpresents a single intense blue-emitting band from 350 nm to 500 nm due to the 5d-4f transition of Eu2+, indicating that α-Sr2P2O7:Eu2+ phosphor powders are suitable for near-UV light-emitting-diode (LED) chips (360-400 nm). The influence of temperature on the luminescence intensity of α-Sr2P2O7:1%Eu2+ is investigated. The activation energy (Ea) for thermal quenching is reported. The phosphor shows excellent thermal stability on temperature quenching. The luminescence properties show that this host material has a highly promising blue-emitting phosphor for white-LED applications.

With the increasing development in the field of electronics, electronic devices have become smaller in size and more heat dissipating. This excessive heat leads to damage to the electronic components, and also their performance becomes bad. Therefore, the process of cooling them must be improved to increase their effectiveness in performance. For this purpose, a numerical study is performed in the seventh chapter to investigate the effect of different nanofluids on heat exchange in a silicon mini channel cooler for cooling electronic components. Three different types of nanofluids are considered (TiO2 -H2O, Ag-H2O, and SWCNT-H2O). In this study, the volumetric fraction of nanoparticles is taken to be 2%, the Reynolds number (Re) is varied between 100 and 700, and the flow regime is assumed to be stationary. The ANSYS Fluent 17.1 commercial software is used as a calculation tool to solve the governing equations, which depend on the finite volume method (FVM) in its solution. The relaxation of decreasing factors used in this study is 0.7 and 0.3 to maintain momentum and pressure, respectively. The residual values of the continuity equation and velocity components are in the range of 10^{-5} and 10^{-6}, respectively, and the second-order upwind scheme has been used. The obtained results confirm that the maximum temperature of the electronic component decreases with the increase in the Reynolds number. The reduction in the temperature of the electronic component is more noticeable for the TiO2-water and SWCNT water nanofluid. Since the values of the coefficient of heat exchange between the channel walls and the nanofluid that contains the single-walled carbon nanotubes nanoparticles are the highest compared to the nanoparticles that do not contain carbon in their composition, therefore, this condition can be considered the best in heat transfer. Therefore, it is recommended that nanofluids containing nanoparticles SWCNT for cooling high-temperature electronic components should be used.

In the eighth chapter, a novel additive based on alkylphenol ethoxylate sulphite is investigated in Zn-Mn electrodeposition on steel from a chloride bath. Electrochemical study via cyclic voltammetry shows that the tested additive increases the over-potential of the Zn deposition, resulting from strong adsorption of molecules additives on the cathode surface. Thus, Mn-rich alloy containing 16.3% of Mn is successfully co-deposited. The morphology and crystallographic structure of Zn-Mn co-deposits are analyzed using Scanning Electron Microscopy (SEM) and X-Ray Diffraction (XRD), respectively. SEM micrographs show that Zn-16.3% Mn alloy obtained in the presence of the tested additive displays hexagonal pyramid morphology. XRD analysis exhibited that Zn-16.3% Mn alloy is monophasic with hexagonal close-packed ε-Zn-Mn phase.

The purpose of the ninth chapter is to empower the scientific and technological community with the knowledge to identify and define key concepts of fire modeling and to develop the ability to apply the CFD (Computer Fluid Dynamics) tools to fire investigation and prevention using basic mathematical models. Combustion, thermal radiation, turbulence, fluid dynamics, and other physical and chemical processes all contribute to the complexity of fire processes. Flame shape, plume behavior, combustion product diffusion, and thermal radiation effects on neighboring objects can all be modeled with Large Eddy Simulation (LES) software. This paper uses many small and large-scale case studies under various boundary conditions to demonstrate the strength of the Fire Dynamics Simulator (FDS), an LES code, established by the National Institute of Standards and Technology (NIST).

Wind energy is renewable energy that does not require any fuel, does not create greenhouse gases, and does not produce toxic or radioactive waste. Wind power offers the possibility of reducing the operating costs of the electricity system. Vertical axis wind turbines (VAWT) of the Darrieus type, especially in small installations, are increasingly appreciated in current research on wind energy. H-shaped turbines may provide appealing spaces for new design strategies that seek to reduce the visual effect of the rotors and then boost their degree of integration in a variety of installation contexts. The main purpose of the tenth chapter is to define and critically evaluate the main design parameters of a 10 kW H-Darrieus vertical axis wind turbine that can be considered a candidate for rural and off-grid urban applications.

Zied Driss
National School of Engineers of Sfax (ENIS)
Laboratory of Electro-Mechanic Systems (LASEM)
University of Sfax
Sfax
Tunisia

List of Contributors

Aïcha Mbarek   Advanced Materials Laboratory, National School of Engineers of Sfax, University of Sfax, BP W 3038, Sfax, Tunisia
Blaise Pascal University, Institute of Chemistry of Clermont-Ferrand, UMR 6296 CNRS, BP 10448, 63000 Clermont-Ferrand, France

Aymen Mohamed Kethiri   Laboratory of Materials and Energy Engineering, University of Mohamed Khider Biskra, Algeria

Belhi Guerira   Laboratory of Mechanical Engineering, University of Mohamed Khider Biskra, Algeria

Ines Ben Naceur   National Engineering School of Sfax, Department of Materials Engineering and Environment(LGME) ENIS, B.P.W.1173 Sfax, Tunisia

Blidi Djamel   Department of Mechanical Engineering, Laboratory of Structures Mechanics and Solids LMSS, Faculty of Technology, University Djillali Liabes Sidi Bel Abbes 22000, Algeria

Bouderne Hamid   Department of Mechanical Engineering, Laboratory of Structures Mechanics and Solids LMSS, Faculty of Technology, University Djillali Liabes Sidi Bel Abbes 22000, Algeria

Fateh Ferroudji   Research Unit in Renewable Energy in Saharan Medium, Road of Reggane –Adra, Algeria

Kamel Chadi   Laboratory of Materials and Energy Engineering, University of Mohamed Khider Biskra, Algeria

Khaled Elleuch   National Engineering School of Sfax, Department of Materials Engineering and Environment(LGME) ENIS, B.P.W.1173 Sfax, Tunisia

Mabrouk Mosbahi   National School of Engineers of Sfax (ENIS), Laboratory of Electro-Mechanic Systems (LASEM), University of Sfax, Sfax, Tunisia

Mariem Lajnef   National School of Engineers of Sfax (ENIS), Laboratory of Electro-Mechanic Systems (LASEM), University of Sfax, Sfax, Tunisia

Mongi Feki   Laboratory of Material Engineering and Environment, ENIS-Tunisia, University of Sfax, Sfax, Tunisia

Mouna Derbel   Advanced Materials Laboratory, National School of Engineers of Sfax, University of Sfax, BP W 3038, Sfax, Tunisia

Miloua Hadj   Department of Mechanical Engineering, Laboratory of Structures Mechanics and Solids LMSS, Faculty of Technology, University Djillali Liabes Sidi Bel Abbes 22000, Algeria

Nourredine Belghar   Laboratory of Materials and Energy Engineering, University of Mohamed Khider Biskra, Algeria

Nouha Loukil   Laboratory of Material Engineering and Environment, ENIS-Tunisia, University of Sfax, Sfax, Tunisia

Nora Boultif   Laboratory of Materials and Energy Engineering, University of Mohamed Khider Biskra, Algeria

Sobhi Frikha   Laboratory of ElectroMechanical Systems (LASEM), National Engineering School of Sfax (ENIS), University of Sfax, 3038 Sfax, Tunisia

Soummar Ahmed   Department of Mechanical Engineering, Laboratory of Structures Mechanics and Solids LMSS, Faculty of Technology, University Djillali Liabes Sidi Bel Abbes 22000, Algeria

Soumia Benbouta   Laboratory of Mechanics of Structures and Materials, Department of Mechanical Engineering, Faculty of Technology, University of Batna 2, Algeria

Toufik Ouattas   Laboratory of Mechanics of Structures and Materials, Department of Mechanical Engineering, Faculty of Technology, University of Batna 2, Algeria

Zied Driss   National School of Engineers of Sfax (ENIS), Laboratory of Electro-Mechanic Systems (LASEM), University of Sfax, Sfax, Tunisia

CHAPTER 1

Numerical Study of a Hydrokinetic Turbine

Mabrouk Mosbahi[1,*], Mariem Lajnef[1] and **Zied Driss[1]**

[1] National School of Engineers of Sfax (ENIS), Laboratory of Electro-Mechanic Systems (LASEM), University of Sfax, Sfax, Tunisia

Abstract: Twisted Darrieus turbine was suggested as an amelioration of conventional Darrieus rotor by modifying it to have helical blades. This reform affords the twisted turbine better performances with regard to the conventional turbine. In this chapter, a computational study of a twisted Darrieus rotor was conducted through the unsteady Reynolds-Averaged Navier-Stokes (URANS) equations. Different grid sizes were investigated to assess the impact of grid generation on the computing findings. The validation of the computing method with antecedent tests was carried out to select the adequate grid size. The flow characteristics of the water around the twisted Darrieus rotor have been assessed and discussed.

Keywords: CFD, Grid generation, Numerical simulations, Twisted blades, Unsteady state, URANS.

INTRODUCTION

Nowadays, the utilization of sustainable energy sources is necessary to lower greenhouse gas emissions in the atmosphere [1 - 4]. Among these green energy sources, hydropower is a sustainable energy source that might be developed in the future. Even though it can not fully substitute the non-renewable sources of energy, hydropower can be an interesting and green substitute [5 - 8]. Thus, it is required to investigate the hydrokinetic turbine design to produce electricity from the water current. The water turbines can be classified into two major kinds; the axial-flow rotors (AFR) and the cross-flow rotors (CFR). The simplicity of the blade shapes and the independence of the water current direction give the advantage to the CFR for the generation of small-scale hydropower with regard to the AFR. Because of the rising cost incurred in the experimental investigations, researchers have used CFD. (Computational Fluid Dynamics) and analytical methods [9 - 11]. The CFD procedure provides the ability to assess the characteristics of fluid flow around a hydraulic turbine that is hard to be assessed

---

* **Corresponding author Mabrouk Mosbahi:** National School of Engineers of Sfax (ENIS), Laboratory of Electro-Mechanic Systems (LASEM), University of Sfax, Sfax, Tunisia; E-mail: mabrouk.mosbahi@gmail.com

using experimental methods. In this context, Moghimi and Motawej [12] carried out a computational test of a twisted Darrieus water rotor (TDWR). They investigated the impact of the twist angle on the operational parameters of the TDWR. In conclusion, the lowest coefficient of power value was obtained with a 120° twist angle. However, the peak one was recorded with a 30° twist angle at a tip-speed ratio value of 3.5. Bianchini *et al.* [13] carried out two-dimensional (2D) CFD investigations of the Darrieus water turbine. In conclusion, they confirmed that a 2D investigation gives the possibility to predict the performance parameters of the turbines with high accuracy and to visualize the flow characteristics around the rotor blades with moderate computational cost. Based on the FLUENT solver, Elbatran *et al.* [14] investigated a hydraulic turbine without and with deflector system at of. In conclusion, they confirmed that the value of 0.4375 was the optimal diameter ratio of the deflector system. Moreover, they affirmed that the performance of the hydraulic rotor could be risen by 78% using a ducted nozzle. The peak value of the coefficient of power reached 0.25 at a TSR of 0.73. Gorle *et al.* [15] computationally and experimentally tested a Darrieus water turbine. They analyzed the field of the fluid flow in the vicinity of the rotor and the performance parameters of the Darrieus rotor. Sarma *et al.* [16] investigated computationally and experimentally a Savonius rotor. They adopted FLUENT software to assess the operational parameters of the turbine for feeble speed boundary. Derakhshan *et al.* [17] conducted computational and experimental tests of a novel CFR. In conclusion, adequate operational parameters were obtained for area with height ratios and for a distance of 13×D between neighbor turbines in a four turbine farm. Using Ansys CFX, Marsh *et al.* [18] studied the effect of two and three-dimension domain selection and the turbulence model on the performance characteristics of CFR. They confirmed that the use of three-dimension domain and k-ω SST model with a boundary layer meshes near the rotor vanes provides accurate computational results. Thakur *et al.* [19] tested numerically a hydraulic turbine with and without an impinging jet duct design. In conclusion, the proposed configuration improves the operational parameters of the hydraulic turbine. The peak value of coefficient of power reached 0.35 at TSR of 0.64 for a conventional turbine. Nevertheless, it reached 0.5 at TSR of 0.61 using the proposed design. Fertahi *et al.* [20] conducted computing investigations on Savonius-Darrieus rotor. The influence of the rotor speed direction on the performance parameters of the hybrid rotor was assessed. They noted that the hybrid turbine with identical rotor speed direction for Savonius and Darrieus turbines outperformed the other hybrid-studied designs. Liang *et al.* [21] studied a combined Darrieus-Savonius rotor. Computing investigations were performed using the URANS equations. The tested Darrieus turbine presented a NACA 0012 profile with a chord of 220 mm. Two-semicircle vanes with an overlap distance of 0.1 characterized the Savonius rotor. They affirmed that the attachment angle, the

Darrieus turbine vanes number and the radius ratio effected the performance parameters of the hybrid rotor. The optimal design for the combined turbine presented a two bladed Darrieus turbine, a radius ratio of 0.25 and an attachment angle of 0°. The peak value of the power coefficient (PC) of the optimal design reached 0.363.

As mentioned in previous works, researchers are using computational techniques to study the CFR. Thus, the computing boundaries selection, *i.e.*, the grid size, is required. In that regard, this chapter focuses on grid size influence on performance parameters of a twisted Darrieus water rotor. The choice of the apt grid size was based on anterior experimental outcomes.

Physical Model

Fig. (**1**) illustrates the TDWR, which presents a design akin to the one tested by the Renewable Energy Center at the New Hampshire University. In the Renewable Energy Center, Bachant *et al.* [22] investigated experimentally a TDWR, which was fabricated by Lucid Energy Company. The TDWR was studied in a testbed for CFR. The testing site presents 37 m of length, 4 m of width and 3 m of depth. The details of the rotor construction are illustrated in Table **1**.

Table 1. Geometric parameters of the TDWR.

Parameter	Value
TDWR diameter (D)	1 m
TDWR height (H)	1.32 m
Number of blades	3
Blade overlap	0.5
solidity (σ)	0.14
Blade profile	NACA 0020

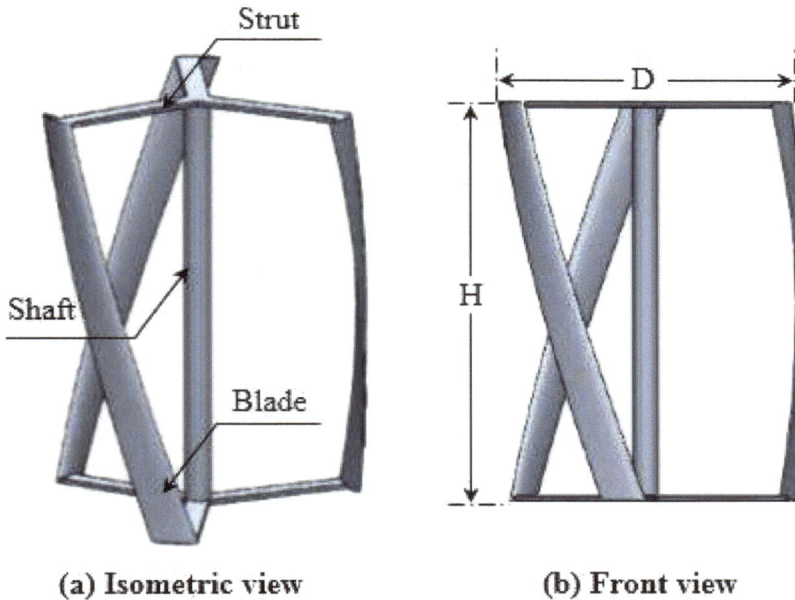

(a) Isometric view   (b) Front view

Fig. (1). Schematic of TDWR.

Numerical Method

Using ANSYS-FLUENT, 3D numerical investigations of TDWR were performed. This code resolves the Navier-Stokes (NS) equations with a finite volume discretization method.

Computing Domain and Limit Conditions

Fig. (**2**) describes the considered computing domain and limit conditions. With the use of the ANSYS Design Modeler, the computing domain is split into two sub-domains (fixed subdomain and rotor subdomain) segregated by a sliding interface. The interface condition is used to ensure the continuity of the flow field and allows the rotation of a domain while the other is kept steady. The fixed subdomain has a rectangular form with 25 m of length, 4 m of width and 3 m of height. The rotor subdomain, which encloses the TDWR, has a cylindrical form. It is characterized by 1.2 m in diameter. For the limit conditions, a velocity of 0.9 m.s^{-1} is defined as an inlet upriver of the TDWR. Downriver the TDWR, atmospheric pressure is defined as an outlet. The rotating shaft of the TDWR considered in this work is the z-axis. For the side and bottom walls of the fixed subdomain, a slip limit condition is put. To the top of the fixed subdomain, a symmetrical limit condition is put. Moreover, a rotating wall with the no-slip condition is put on the TDWR blades. The rotational speed of the rotating subdomain varies depending on TSR. Thus, various numerical investigations are conducted to have computing findings at the designated TSR.

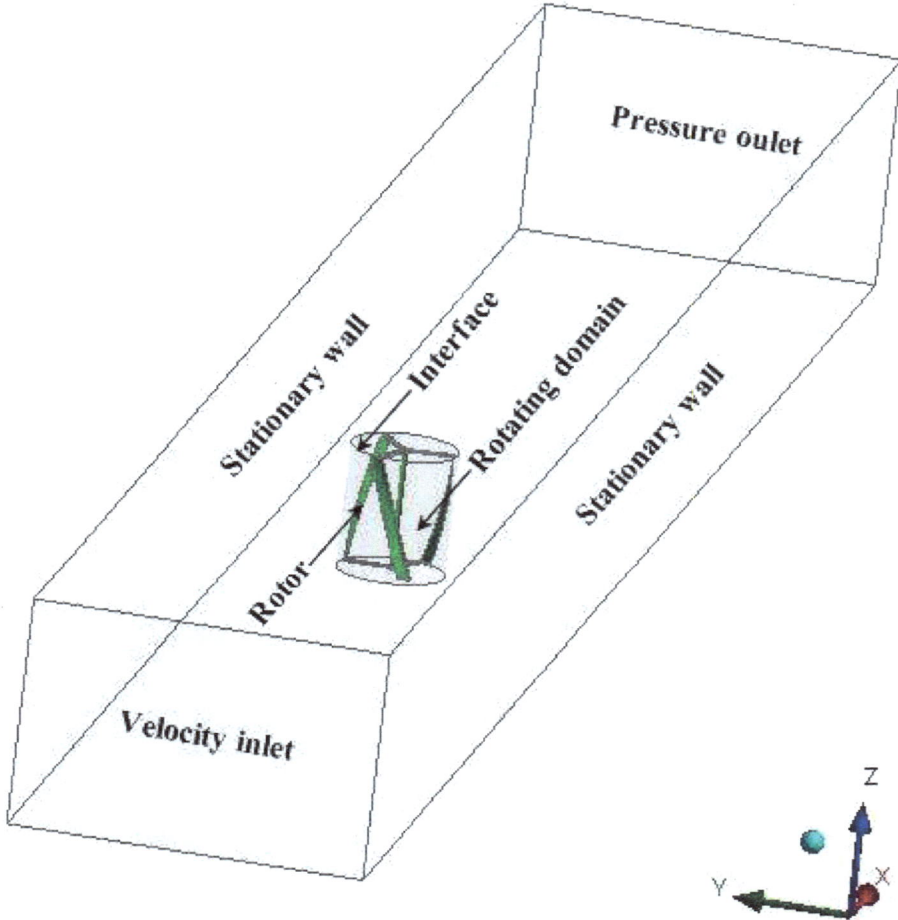

Fig. (2). Limit conditions.

Meshing

To ensure proper numerical findings, the selection of the best grid size is necessary. In fact, a non-compliant unorganized grid with elements, which present tetrahedral forms, is considered in this study owing to the complex geometry of the TDWR. In this part, various grid sizes are tested to achieve grid independent findings.

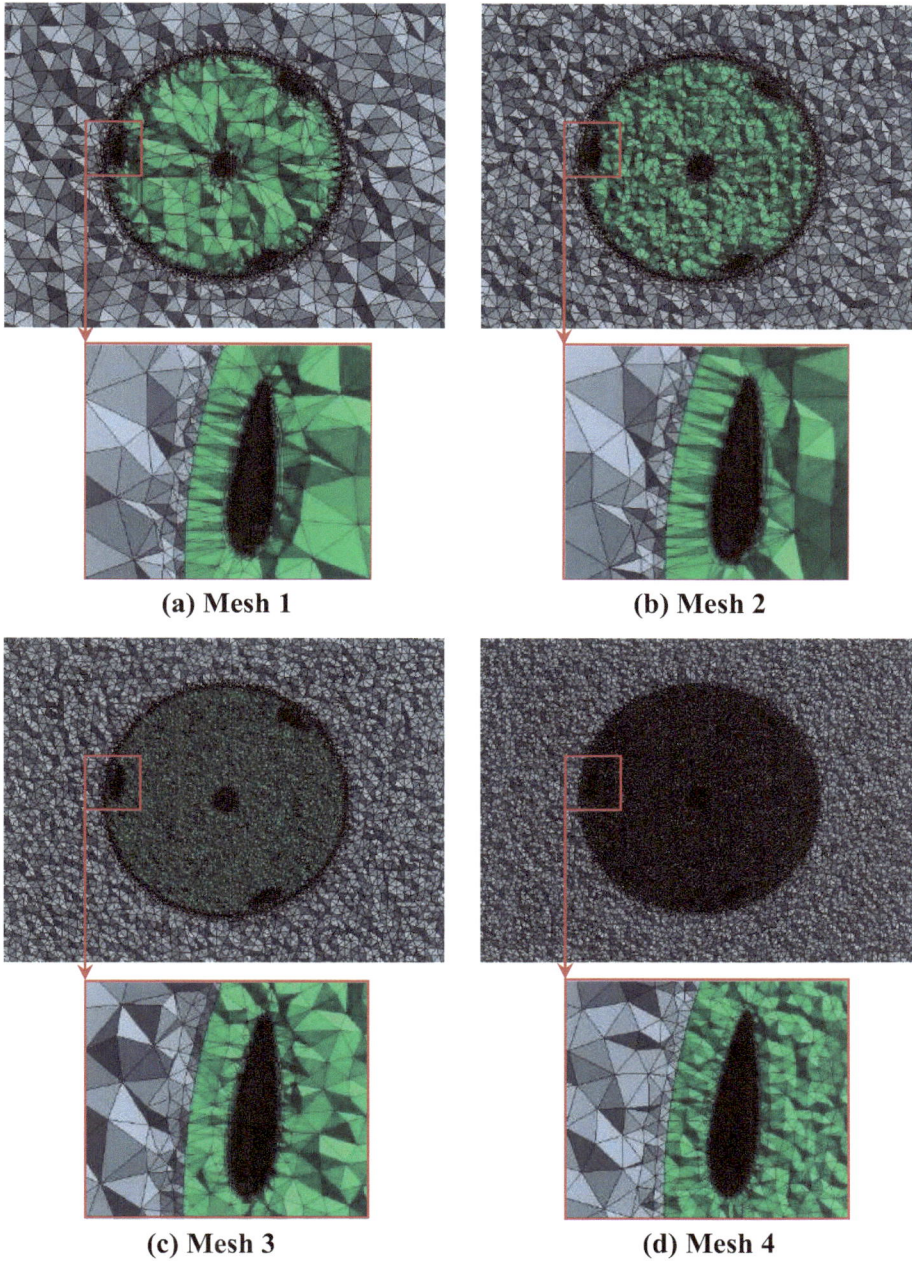

(a) Mesh 1   (b) Mesh 2

(c) Mesh 3   (d) Mesh 4

Fig. (3). CFD mesh.

The selection of the suitable grid size is based on the comparison between our computational findings and the experimental results founded by Bachant *et al.*

[22]. The different grid kits present a number of elements of 11 million (Mesh 1), 14 million (Mesh 2), 17 million (Mesh 3) and 19 million (Mesh 4). As it is illustrated in (Fig. **3**), the fine grid is created in the rotor subdomain compared to the stationary subdomain. To capture the quick changes of the aerodynamic characteristics around the TDWR, it is advisable to create a fine grid near the TDWR walls sufficiently: boundary layer mesh. Thus, a prismatic layer is created at the level of boundary layers and especially at the TDWR blades.

Numerical Settings

In this study, a FLUENT solver is utilized to resolve the unsteady incompressible RANS equations. The NS equations according to a Newtonian fluid are prescribed under the continuity, momentum and turbulence model (TM) equations.

Continuity equation:

$$\frac{\partial \rho}{\partial t} + \frac{\partial (\rho \, u_i)}{\partial x_i} = 0 \tag{1}$$

Momentum equation:

$$\frac{\partial (\rho \, u_i)}{\partial t} + \frac{\partial (\rho \, u_i u_j)}{\partial x_j} = -\frac{\partial p}{\partial x_i} + \frac{\partial}{\partial x_j}\left[\mu \left(\frac{\partial u_j}{\partial x_j} + \frac{\partial u_j}{\partial x_i} - \frac{2}{3}\delta_{ij}\frac{\partial u_i}{\partial x_i} \right) \right] +$$
$$\frac{\partial (-\rho \, \overline{u_i' u_j'})}{\partial x_j} + F_i \tag{2}$$

Where:

$$-\rho \overline{u_i' u_j'} = \mu_t \left(\frac{\partial u_i}{\partial x_j} + \frac{\partial u_j}{\partial x_i} \right) - \frac{2}{3}\rho \, k \, \delta_{ij} \tag{3}$$

Where:

$\overline{u_i' u_j'}$:Turbulent stress, p: Pressure, ρ: Density of the water and F_i : External forces.

To resolve the NS equations, a TM is required. Numerous TMs are available in FLUENT solver. The most used TM applied in previous works is the realizable k-ε TM. In fact, previous works [23 - 25] confirmed that the realizable k-ε TM is the most adequate to simulate the flow through turbomachinery and separated flows or boundary layer. The realizable k-ε model is defined by transport equations and a novel term for the turbulent viscosity, including a variable C_μ calculated using the following equation:

$$C_\mu = \frac{1}{A_0 + A_S \dfrac{kU^*}{\varepsilon}} \tag{4}$$

Where:

$$U^* = \sqrt{S_{ij}S_{ij} + \tilde{\Omega}_{ij}\tilde{\Omega}_{ij}} \tag{5}$$

$$S_{ij} = \frac{1}{2}(\frac{\partial \mu_j}{\partial x_i} + \frac{\partial \mu_i}{\partial x_j}) \tag{6}$$

$$\tilde{\Omega}_{ij} = \Omega_{ij} - 2\varepsilon_{ijk}\omega_k \tag{7}$$

$$\Omega_{ij} = \bar{\Omega}_{ij} - 2\varepsilon_{ijk}\omega_k \tag{8}$$

The transport equations for k and ε in the realizable k- ε TM can be written as follows:

$$\frac{\partial}{\partial t}(\rho\, k) + \frac{\partial}{\partial x_j}(\rho\, k\, u_j) = \frac{\partial}{\partial x_j}\left[\left(\mu + \frac{\mu_t}{\sigma_k}\right)\frac{\partial k}{\partial x_j}\right] + G_k + G_b -$$
$$\rho\, \varepsilon - Y_M + S_k \tag{9}$$

$$\frac{\partial}{\partial t}(\rho\, \varepsilon) + \frac{\partial}{\partial x_j}(\rho\, \varepsilon\, u_j) = \frac{\partial}{\partial x_j}\left[\left(\mu + \frac{\mu_t}{\sigma_\varepsilon}\right)\frac{\partial \varepsilon}{\partial x_j}\right] + \rho\, C_1 S_\varepsilon -$$
$$\rho\, C_2 \frac{\varepsilon^2}{k + \sqrt{\upsilon\varepsilon}} + C_{1\varepsilon}\frac{\varepsilon}{k}C_{3\varepsilon}G_b + S_\varepsilon \tag{10}$$

where:

$$C_1 = \max\left[0.43, \frac{\eta}{\eta+5}\right] \tag{11}$$

$$\eta = S\frac{k}{\varepsilon} \tag{12}$$

$$S = \sqrt{2S_{ij}S_{ij}} \tag{13}$$

Where:

u_i: Velocity components, x_j: Cartesian coordinate, t: Time, μ: Viscosity, G_k and G_b: Turbulent kinetic energy generations, σ_k and σ_ε: Turbulent Prandtl numbers, S_k and S_ε: Source terms, C_1 and C_ε: Constants, S: Assumed source terms and Ω_{ij}: Mean rate of rotation tensor viewed in a rotating reference frame with the angular speed ω_k. After considering the realizable k-ε model, a pressure-velocity coupling procedure (Semi-Implicit Method for Pressure-Linked Equations) with second-order upwind scheme for the convective terms is adopted to run simulations. The scaled residuals with the value of 10^{-6} are considered as the convergence criteria for each time step. The time step is defined as 1°/time step, which means in each time step, the rotor turned 1°. The time step size considered is employed with 50 iterations per time step. Computational findings are deemed to be statistically stable when fluctuations of the torque coefficient (TC) reach a quasi-steady state. In this study, four revolutions are needed for each simulation to obtain the quasi-steady state. Fig. (4) presents the torque coefficient variation for four revolutions of the TDWR blade. In the first step, an irregularity was noticed. A periodic solution was obtained after three revolutions.

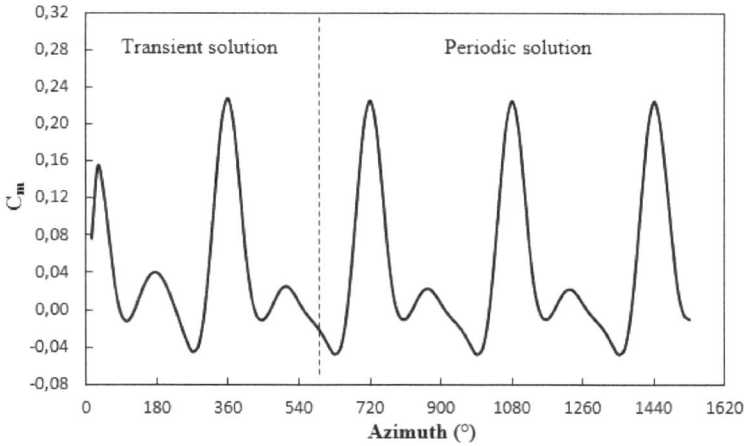

Fig. (4). TC over four cycles of the rotor.

Comparison with Experimental Results

When using limited computing resources, obtaining a solution that does not depend on the increment of the grid elements number is essential to avert the rise of the computing stations need without remarkable modification in outcomes. In this study, mesh independence findings were obtained by modifying the grid elements number in the computing domain, including TDWR. The TC, defined by equation (14), is selected to be seen for the mesh independence test.

$$C_m = \frac{M}{\frac{1}{2}\rho\, V_\infty^{2}\, R\, S} \tag{14}$$

Where:

M: TDWR torque,

R: TDWR radius,

S: Frontal area of the TDWR.

$$S = 2RH \tag{15}$$

Where:

H: TDWR height.

Fig. (**5**) presents the influence of the increment of the mesh element number on the TDWR TC at TSR of 2. From these outcomes, it is confirmed that and provided similar outcomes.

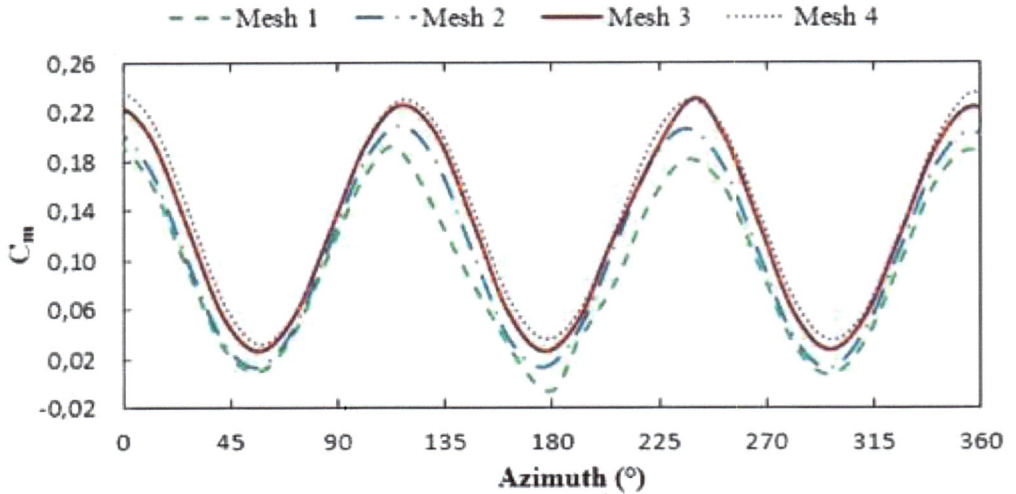

Fig. (5). Torque coefficient.

Fig. (**6**) illustrates a comparison of the PC between computing findings and experimental outcomes at the range of TSR. The PC is defined by the following equation:

$$C_p = \frac{P}{\frac{1}{2}\rho\,SV_\infty^{\,3}} = \lambda C_m \tag{16}$$

Where:

P: Generated power,

λ: TSR.

$$\lambda = \frac{\omega R}{V_\infty} \tag{17}$$

Where:

ω:Rotational speed.

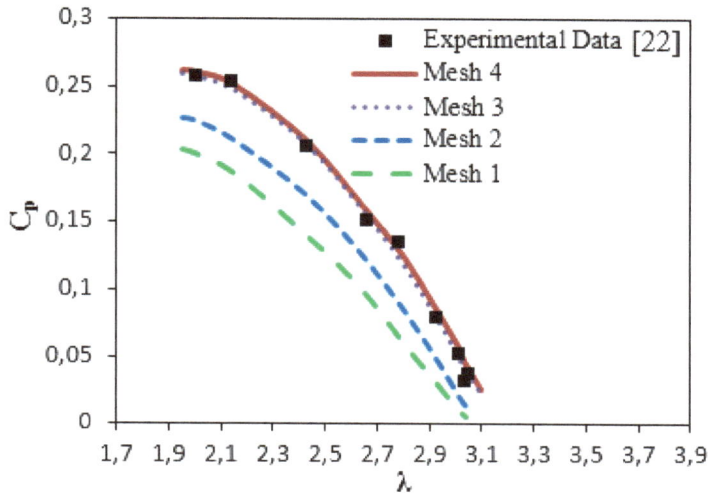

Fig. (6). Power coefficient.

From these outcomes, it is confirmed that Mesh 3 and Mesh 4 are in good agreement with the experimental findings. The error compared to the experimental curve of the PC is more remarkable for Mesh 1 and Mesh 2. Mesh 3 is chosen for more precision and time-saving with an average error of 5% compared to experimental findings.

RESULTS AND DISCUSSION

Magnitude Velocity

Fig. (7) illustrates the contour plots of water velocity (WV) at TSR of 2 round TDWR for the different grid sizes.

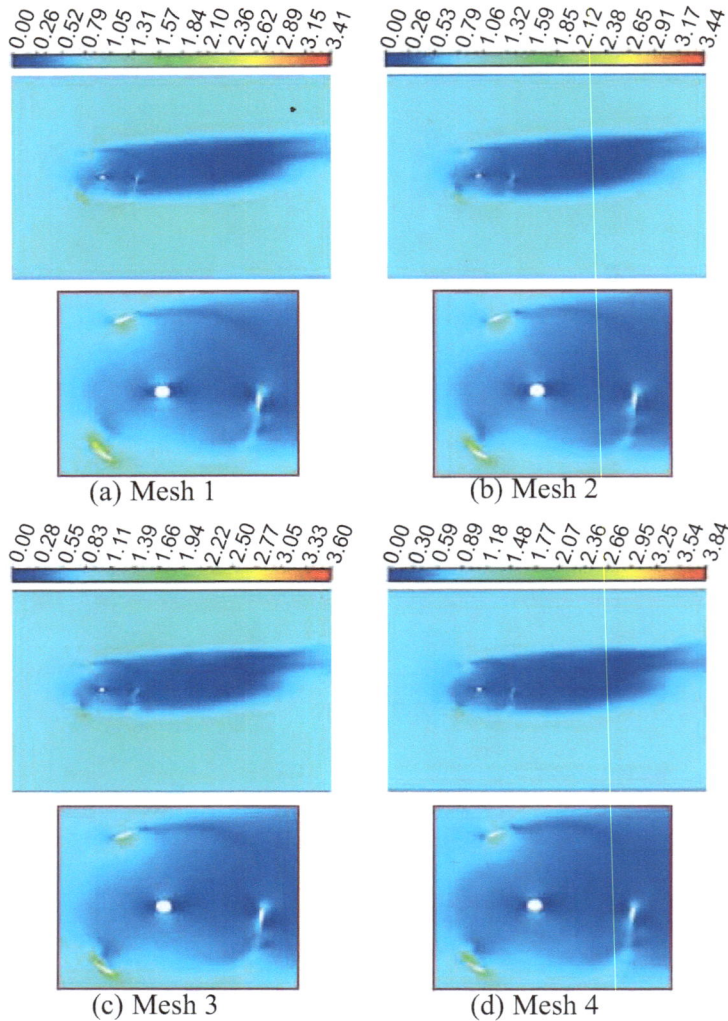

(a) Mesh 1

(b) Mesh 2

(c) Mesh 3

(d) Mesh 4

Fig. (7). Contour plots of the water velocity.

According to these outcomes, the WV plots are identical for the various grid element numbers. In fact, the WV is feeble upriver the TDWT. It is the same as the value as put at the inlet limit condition of the stationary subdomain. The TDWR downriver presents a large wake area that displays the mark of a feeble WV. In addition, it has been observed that the slowing area deviates up slightly. This fact is a result of the forces bred by the vanes of the TDWR owing to the TDWR rotation anticlockwise. By looking closely at the area encircled the

TDWR, a water quickening area is seen on the intake side, nearly the foremost edge of the TDWR blade. Thereby, a speeding area on the blade intake side is developed. While varying the grid element number, a variance in values of the WV is spotted. The summit value of WV is equal to 3.41 m.s^{-1} using Mesh 1, to 3.44 m.s^{-1} using Mesh 2, to 3.60 m.s^{-1} using Mesh 3 and to 3.84 m.s^{-1} using Mesh 4. Therefore, by comparing these outcomes, it could be proved that changing the grid size influences the contour plots of the WV around the TDWR.

Static Pressure

Fig. (**8**) illustrates the contour plots of the static pressure (SP) around TDWR for various grid element numbers at TSR of 2. From these outcomes, it could be seen that the SP is identical for the various grid sizes. It is uniform at the inlet of the domain. Large SP is seen upriver the TDWR, whilst low SP is seen downriver the TDWR, which resulted in the variance of SP. Power is developed at that point owing to this variance of SP, and the TDWR commence turning. By comparing these outcomes, it could be proved that the changing of the grid element number influences the contour plots of the SP around the TDWR. The summit value of the SP is equal to 2727 Pa using Mesh 1, to 2812 Pa using Mesh 2, to 2858 Pa using Mesh 3, and to 2881 Pa using Mesh 4.

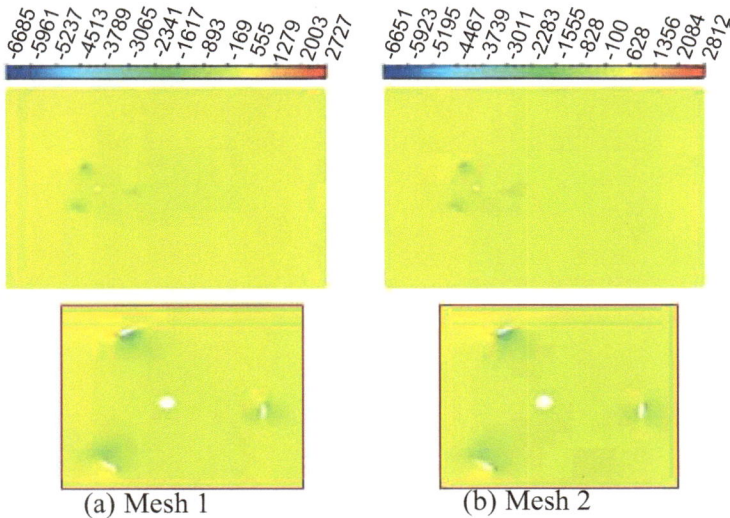

(a) Mesh 1   (b) Mesh 2

Fig. 8 cont.....

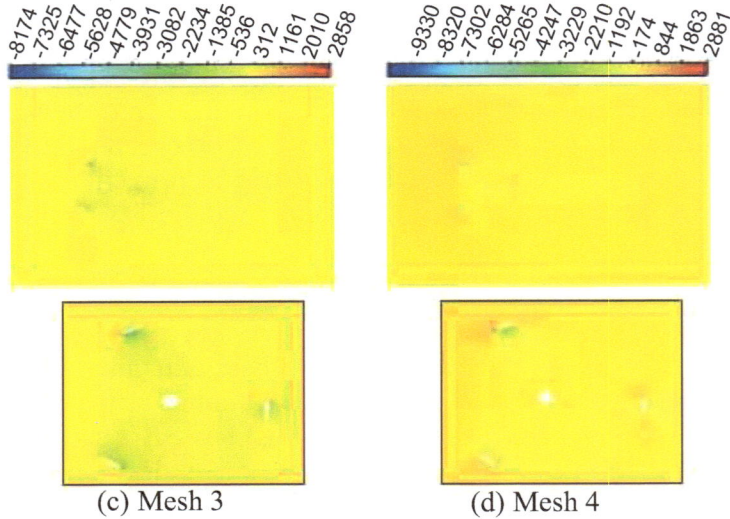

(c) Mesh 3   (d) Mesh 4

Fig. (8). Contour plots of the static pressure.

Turbulent Kinetic Energy

Fig. (**9**) illustrates the contour plots of turbulent kinetic energy (TKE) at TSR of 2 around TDWR for different grid sizes. From the numerical outcomes, it is outstanding that the TKE is quite feeble in the stationary subdomain, excepting the space encircling the TDWR for the various grid sizes. The TKE variations show a rise of values downriver the TDWR. By comparing those outcomes, it could be affirmed that the grid size influences the contour plot of the TKE around the TDWR. The summit value of the TKE is equal to 0.23 $m^2.s^{-2}$ using Mesh 1, to 0.32 $m^2.s^{-2}$ using Mesh 2, to 0.44 $m^2.s^{-2}$ using Mesh 3 and to 0.48 $m^2.s^{-2}$ using Mesh 4.

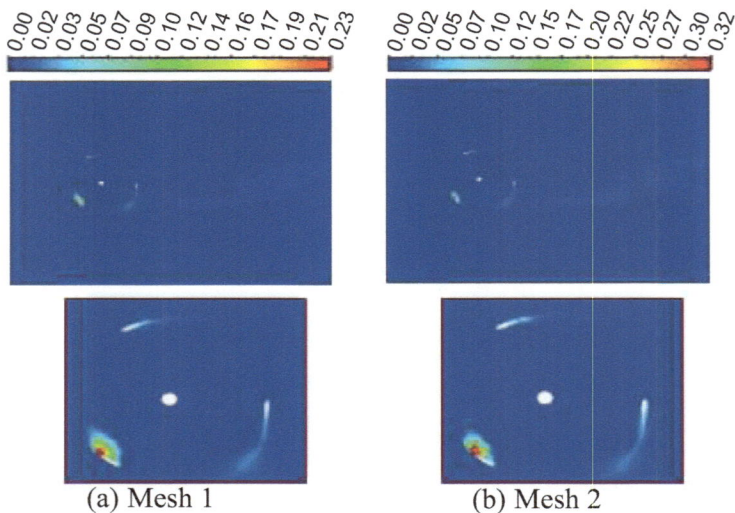

(a) Mesh 1   (b) Mesh 2

Fig. 9 cont.....

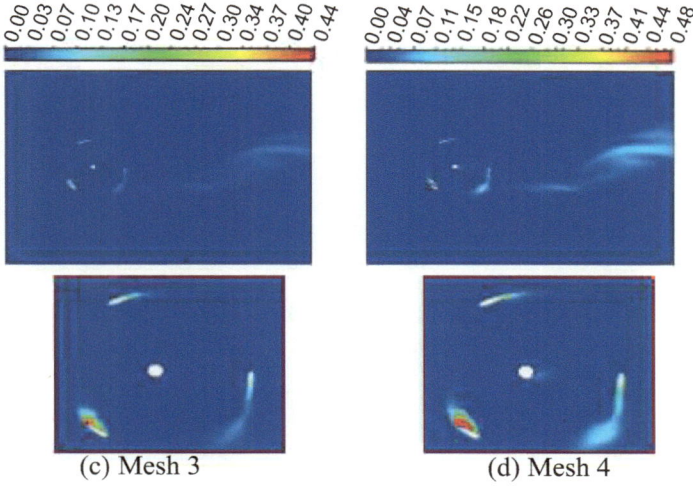

(c) Mesh 3   (d) Mesh 4

Fig. (9). Contour plots of turbulent kinetic energy.

Turbulence Eddy Dissipation

Fig. (**10**) illustrates the contour plots of turbulence eddy dissipation (TED) around the TDWR at TSR of 2 for the different grid sizes. As per computational outcomes, it could be affirmed that an increment of the grid element numbers influences the contour plots of the TED. By comparing the obtained outcomes, it could be affirmed that the TED is quite feeble in the stationary subdomain excepting the space encircling the TDWR. Nevertheless, a rise of the TED is noticed in the vicinity of the TDWR and downriver the TDWR.

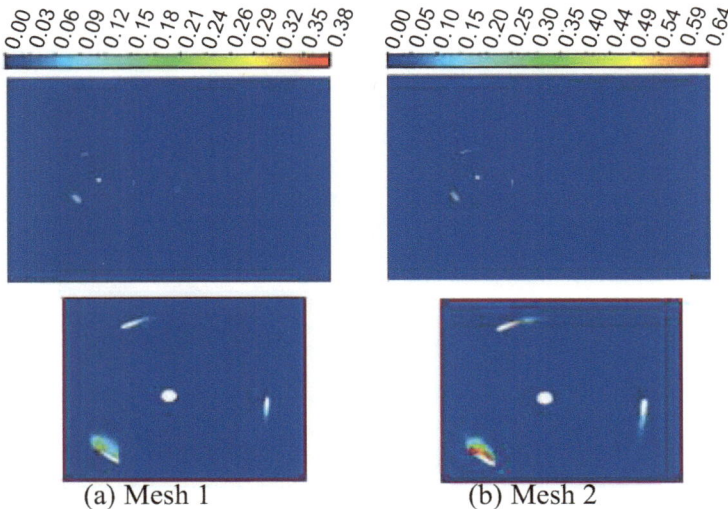

(a) Mesh 1   (b) Mesh 2

Fig. 10 cont.....

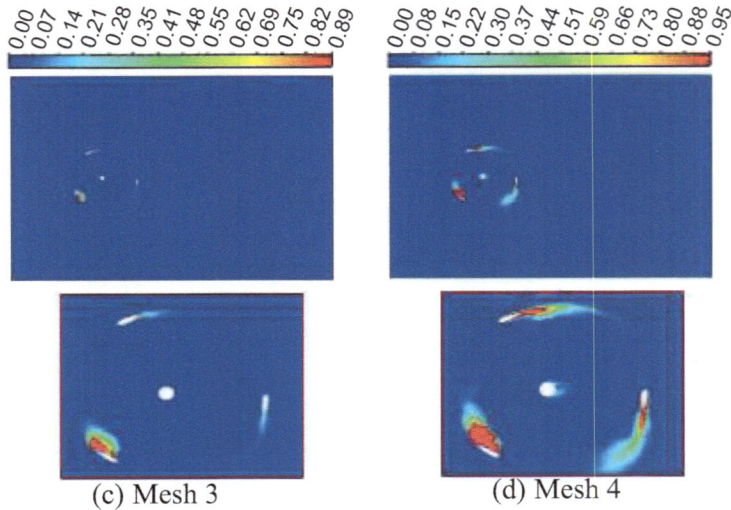

(c) Mesh 3   (d) Mesh 4

Fig. (10). Contour plots of turbulence eddy dissipation.

A dissimilarity in TED values is remarked by changing the grid element numbers. In fact, the peak value of the TED around the TDWR is equal to 0.38 $m^2.s^{-3}$ using Mesh 1, to 0.64 $m^2.s^{-3}$ using Mesh 2, to 0.89 $m^2.s^{-3}$ using Mesh 3 and to 0.95 $m^2.s^{-3}$ using Mesh 4.

Eddy Viscosity

Fig. (**11**) shows the contour plots of eddy viscosity (EV) for dissimilar meshing sizes at TSR of 2 round TDWR.

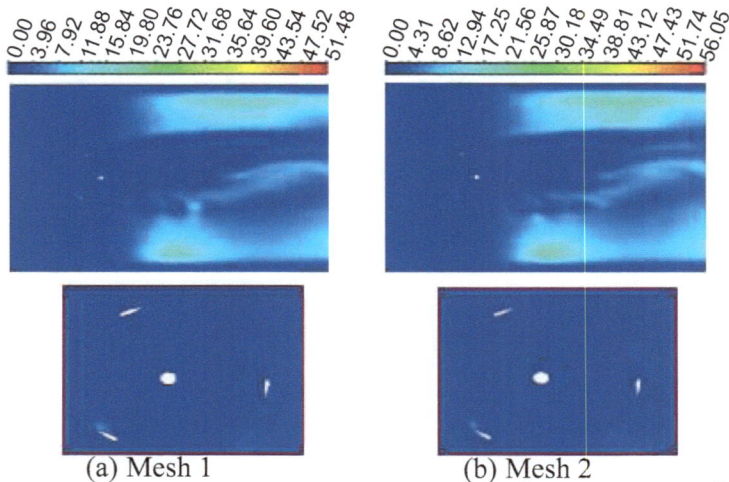

(a) Mesh 1   (b) Mesh 2

Fig. 11 cont.....

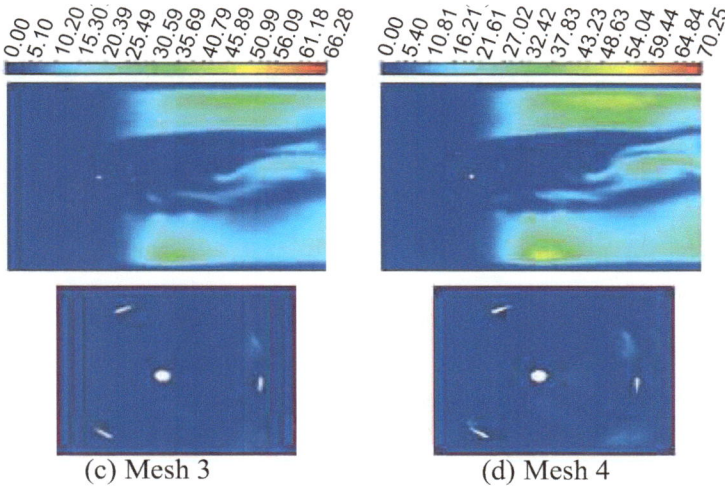

Fig. (11). Contour plots of eddy viscosity.

According to those results, it could be affirmed that EV is the same for the dissimilar grid sizes. In effect, it is feeble upriver the TDWR. It rises around the vanes of the TDWR, and supreme EV values are seen downriver the TDWR. By comparing those outcomes, it could be affirmed that the grid size influences the contour plot of the EV. In fact, the supreme EV value is equal to 51.48 kg.m^{-1}.s^{-1} using the Mesh 1, to 56.05 kg.m^{-1}.s^{-1} using the Mesh 2, to 66.28 kg.m^{-1}.s^{-1} using the Mesh 3 and to 70.25 kg.m^{-1}.s^{-1} using the Mesh 4.

CONCLUDING REMARKS

Computational investigations were conducted to test the grid size influence on the computational outcomes under the present chapter. The computational tests were based on a URANS solver. The validation of the computational model was made using preceding experimental outcomes founded in the literature. Computational outcomes affirmed that grid size influences computational outcomes. Indeed, the effect of the meshing size on the flow characteristics around TDWR has been presented. For our future computational simulations, a meshing with 16.53 million elements is selected to test the TDWR because it provides a great bon-accord with the experimental outcomes and decreases the resolution computation time.

NOMENCLATURE

u_i	Components of the velocity	$(m.s^{-1})$
u_i'	Fluctuating velocity components	$(m.s^{-1})$
V_∞	Velocity of the fluid	$(m.s^{-1})$
w	Width of the stationary subdomain	(m)
x_i	Cartesian coordinate	(m)
y^+	Non dimensional parameter	
ε	Turbulence eddy dissipation	$(W.kg^{-1})$
μ	Dynamic viscosity	$(Pa.s)$
μ_t	Turbulent viscosity	$(Pa.s)$
ρ	Density	$(kg.m^{-3})$
ω	Revolution speed	$(rad.s^{-1})$
λ	TSR	
σ_k	Constant of the k-ε turbulence model	
σ_ε	Constant of the k-ε turbulence model	
δ_{ij}	Chroneker indices	

CONSENT FOR PUBLICATION

Not applicable.

CONFLICT OF INTEREST

The author declares no conflict of interest, financial or otherwise.

ACKNOWLEDGEMENTS

The authors would like to thank the Laboratory of Electro Mechanic Systems' (LASEM) members for their financial assistance.

REFERENCES

[1]   G. Ferrari, D. Federici, P. Schito, F. Inzoli, and R. Mereu, "CFD study of Savonius wind turbine: 3D model validation and parametric analysis", *Renew. Energy,* vol. 105, pp. 722-734, 2017. [http://dx.doi.org/10.1016/j.renene.2016.12.077]

[2]   B.A. Bhayo, and H.H. Al-Kayiem, "Experimental characterization and comparison of performance

parameters of S-rotors for standalone wind power system", *Energy,* vol. 138, pp. 752-763, 2017.
[http://dx.doi.org/10.1016/j.energy.2017.07.128]

[3] D. Petković, N.T. Pavlović, and Ž. Ćojbašić, "Wind farm efficiency by adaptive neuro-fuzzy strategy", *Int. J. Electr. Power Energy Syst.,* vol. 81, pp. 215-221, 2016.
[http://dx.doi.org/10.1016/j.ijepes.2016.02.020]

[4] V. Nikolić, V.V. Mitić, L. Kocić, and D. Petković, "Wind speed parameters sensitivity analysis based on fractals and neuro-fuzzy selection technique", *Knowl. Inf. Syst.,* vol. 52, pp. 255-265, 2017.
[http://dx.doi.org/10.1007/s10115-016-1006-0]

[5] D. Petković, V. Nikolić, V.V. Mitić, and L. Kocić, "Estimation of fractal representation of wind speed fluctuation by artificial neural network with different training algorothms", *Flow Meas. Instrum.,* vol. 54, pp. 172-176, 2017.
[http://dx.doi.org/10.1016/j.flowmeasinst.2017.01.007]

[6] D. Petković, Ž. Ćojbašić, V. Nikolić, S. Shamshirband, M.L.M. Kiah, N.B. Anuar, and A.W.A. Wahab, "Adaptive neuro-fuzzy maximal power extraction of wind turbine with continuously variable transmission", *Energy,* vol. 64, pp. 868-874, 2014.
[http://dx.doi.org/10.1016/j.energy.2013.10.094]

[7] I. Ostos, I. Ruiz, M. Gajic, W. Gómez, A. Bonilla, and C. Collazos, "A modified novel blade configuration proposal for a more efficient VAWT using CFD tools", *Energy Convers. Manage.,* vol. 180, pp. 733-746, 2019.
[http://dx.doi.org/10.1016/j.enconman.2018.11.025]

[8] M.M. Nunes, R.C.F. Mendes, T.F. Oliveira, and A.C.P.B. Junior, "An experimental study on the diffuser-enhanced propeller hydrokinetic turbines", *Renew. Energy,* vol. 133, pp. 840-848, 2019.
[http://dx.doi.org/10.1016/j.renene.2018.10.056]

[9] M. Milovančević, V. Nikolić, D. Petkovic, L. Vracar, E. Veg, N. Tomic, and S. Jović, "Vibration analyzing in horizontal pumping aggregate by soft computing", *Measurement,* vol. 125, pp. 454-462, 2018.
[http://dx.doi.org/10.1016/j.measurement.2018.04.100]

[10] S. Gavrilović, N. Denić, D. Petković, N.V. Živić, and S. Vujičić, "Statistical evaluation of mathematics lecture performances by soft computing approach", *Comput. Appl. Eng. Educ.,* vol. 26, no. 4, pp. 902-905, 2018.
[http://dx.doi.org/10.1002/cae.21931]

[11] D. Petković, "Prediction of laser welding quality by computational intelligence approaches", *Optik (Stuttg.),* vol. 140, pp. 597-600, 2017.
[http://dx.doi.org/10.1016/j.ijleo.2017.04.088]

[12] M. Moghimi, and H. Motawej, "Developed DMST model for performance analysis and parametric evaluation of Gorlov vertical axis wind turbines", *Sustainable Energy Technologies and Assessments,* vol. 37, p. 100616, 2020.
[http://dx.doi.org/10.1016/j.seta.2019.100616]

[13] A. Bianchini, F. Balduzzi, P. Bachant, G. Ferrara, and L. Ferrari, "Effectiveness of two-dimensional CFD simulations for Darrieus VAWTs: a combined numerical and experimental assessment", *Energy Convers. Manage.,* vol. 136, pp. 318-328, 2017.
[http://dx.doi.org/10.1016/j.enconman.2017.01.026]

[14] A.H. Elbatran, Y.M. Ahmed, and A.S. Shehata, "Performance study of ducted nozzle Savonius water turbine, comparison with conventional Savonius turbine", *Energy,* vol. 134, pp. 566-584, 2017.
[http://dx.doi.org/10.1016/j.energy.2017.06.041]

[15] J.M.R. Gorle, L. Chatellier, F. Pons, and M. Ba, "Flow and performance analysis of H-Darrieus hydro turbine in a confined flow: A computational and experimental study", *J. Fluids Structures,* vol. 66, pp. 382-402, 2016.
[http://dx.doi.org/10.1016/j.jfluidstructs.2016.08.003]

[16]   N.K. Sarma, A. Biswas, and R.D. Misra, "Experimental and computational evaluation of Savonius hydrokinetic turbine for low velocity condition with comparison to Savonius wind turbine at the same input power", *Energy Convers. Manage.*, vol. 83, pp. 88-98, 2014.
[http://dx.doi.org/10.1016/j.enconman.2014.03.070]

[17]   S. Derakhshan, M. Ashoori, and A. Salemi, "Eexperimental and numerical study of a vertical axis tidal turbine performance", *Ocean Eng.*, vol. 137, pp. 59-67, 2017.
[http://dx.doi.org/10.1016/j.oceaneng.2017.03.047]

[18]   P. Marsh, D. Ranmuthugala, I. Penesis, and G. Thomas, "The influence of turbulence model and two and three-dimensional domain selection on the simulated performance characteristics of vertical axis tidal turbines", *Renew. Energy,* vol. 105, pp. 106-116, 2017.
[http://dx.doi.org/10.1016/j.renene.2016.11.063]

[19]   N. Thakur, A. Biswas, Y. Kumar, and M. Basumatary, "CFD analysis of performance improvement of the Savonius water turbine by using an impinging jet duct design", *Chin. J. Chem. Eng.,* vol. 27, no. 4, pp. 794-801, 2019.
[http://dx.doi.org/10.1016/j.cjche.2018.11.014]

[20]   S.D. Fertahi, T. Bouhal, O. Rajad, T. Kousksou, A. Arid, T. El Rhafiki, A. Jamil, and A. Benbassou, "CFD performance enhancement of a low cut-in speed current Vertical Tidal Turbine through the nested hybridization of Savonius and Darrieus", *Energy Convers. Manage.*, vol. 169, pp. 266-278, 2018.
[http://dx.doi.org/10.1016/j.enconman.2018.05.027]

[21]   X. Liang, S. Fu, B. Ou, C. Wu, Y.H.C. Chao, and K. Pi, "A computational study of the effects of the radius ratio and attachment angle on the performance of a Darrieus-Savonius combined wind turbine", *Renew. Energy,* vol. 113, pp. 329-334, 2017.
[http://dx.doi.org/10.1016/j.renene.2017.04.071]

[22]   P. Bachant, and M. Wosnik, "Performance measurements of cylindrical- and spherical-helical cross-flow marine hydrokinetic turbines, with estimates of exergy efficiency", *Renew. Energy,* vol. 74, pp. 318-325, 2015.
[http://dx.doi.org/10.1016/j.renene.2014.07.049]

[23]   A. Kumar, and R.P. Saini, "Performance analysis of a single stage modified Savonius hydrokinetic turbine having twisted blades", *Renew. Energy,* vol. 113, pp. 461-478, 2017.
[http://dx.doi.org/10.1016/j.renene.2017.06.020]

[24]   M. Mosbahi, A. Ayadi, Y. Chouaibi, Z. Driss, and T. Tucciarelli, "Experimental and numerical investigation of the leading edge sweep angle effect on the performance of a delta blades hydrokinetic turbine", *Renew. Energy,* vol. 162, pp. 1087-1103, 2020.
[http://dx.doi.org/10.1016/j.renene.2020.08.105]

[25]   M. Mosbahi, S. Elgasri, M. Lajnef, B. Mosbahi, and Z. Driss, "Performance enhancement of a twisted Savonius hydrokinetic turbine with an upstream deflector", *Int. J. Green Energy,* 2020.
[http://dx.doi.org/10.1080/15435075.2020.1825444]

CHAPTER 2

CFD Simulation of the Heat Transfer using a Cuo-water Nano-fluid in Different Cross-sections of Mini-channels

Kamel Chadi[1,*], **Aymen Mohamed Kethiri**[1], **Nourredine Belghar**[1], **Belhi Guerira**[2] and **Zied Driss**[3]

[1] *Laboratory of Materials and Energy Engineering, University of Mohamed Khider Biskra, Algeria*

[2] *Laboratory of Mechanical Engineering, University of Mohamed Khider Biskra, Algeria*

[3] *Laboratory of Electromechanical Systems (LASEM), National School of Engineers of Sfax (ENIS), University of Sfax (US), B.P. 1173, Road Soukra km 3.5, 3038, Sfax, Tunisia*

Abstract: In the present work, the purpose is to study heat exchange which is directly related to factors such as Reynolds number, thermal properties of materials, geometric shapes and dimensions. A numerical study of the heat exchanges between cross-sections selected of a mini-channel cooler of dimensions ($40 \times 52 \times 6$ mm^3) is carried out. Three different forms have been considered for cooling an electronic component using a nanofluid (CuO-water) as a cooling liquid with 4% volume concentration of nanoparticles. The simulation is carried out using the ANSYS Fluent software. The Reynolds number (*Re*) is taken between 100 and 700 and the stream regime is assumed to be stationary. The results obtained for the three forms of mini-channels proposed show that the raise in the exchange surface between the CuO-water nanofluid and walls of the mini-channels leads to the increase in the heat exchange coefficient and to the amelioration of the maximum temperature of electronic components by increasing the value of the flow velocity. This is confirmed by the results of the third case. In contrast to the first case that does not contain ribs, and the second case, which contains two ribs inside the channel, these two cases provide insufficient heat exchange, and the maximum temperature of the electronic component remains high compared to the third case, which contains four ribs, the latter contribute to the increase in heat exchange inside the channel.

Keywords: ANSYS Fluent, Heat exchange, Minichannels, Nanofluid, Numerical simulation.

---

[*] **Corresponding author Kamel Chadi:** Laboratory of Materials and Energy Engineering, university of Mohamed khider Biskra, Algeria; E-mail: chadikamel_dz@yahoo.fr

Zied Driss (Ed.)

INTRODUCTION

In order to increase the rate of heat transfer, one must take into consideration parameters related to factors, such as Reynolds number, thermal properties of geometric shapes and dimensions. Several studies have been carried out in this area. P. Gunnasegaranet *et al.* [1] conducted a numerical study on the effect of geometric parameters on heat transfer characteristics for three different shapes of microchannels (rectangular, trapezoidal and triangular). They found that the thermal exchange coefficient and the Poiseuille number increases with the increase of Reynolds number (Re). Heat transfer coefficient and Poiseuille number have the highest values in the case of rectangular form. However, for the micro-channels of triangular form, they have the weakest one. The intermediate values have been obtained for the microchannels of trapezoidal form. The results of a numerical simulation [2] using the ANSYS-Fluent termal characteristics software for a mini-channel radiator structure show that the maximum temperature and the termal resistance, decrease with the increase of the speed of the flow. Muhammad Saeed and Man-Hoe [3] present a numerical and experimental investigation on the heat transfer enhancement characteristics using three different volume concentrations of nano-particles Al_2O_3 in water as a base fluid, and with four different cannel configurations of mini-channel heat sink.

They observed the convective heat transfer coefficient of the heat sink with fin spacing. Moreover, it was also observed also that the enhancement factor increases by dreading the fins pacing (hydraulic diameter) of the flow channel at the same value of volume concentration and coolant flow rate.

Yulin *et al.* [4] numerically investigated the convective heat exchanger of water/Al_2O_3 nano-fluid in an inclined square enclosure. The cavity of the upper and lower walls was insulated. Moreover, there is a constant temperature heat source in the center of the enclosure. The enclosure is located under the impact of an inclined magnetic field (MF). In their research, they used finite volume method to solve the governing equations. Their results show that an increase in the angle of the enclosure increases the rate of heat transfer on the right wall by increasing the angle of the magnetic field. Also, the addition of nano-additives results in intensification in the thermal exchange rate.

Asadi *et al.* [5] reviewed the effect of nanoparticle concentration, shape, and temperature on property values , including thermal conductivity, viscosity, density, and specific heat of oil-based nanofluids. Where they concluded that oil-based nanofluids can be efficient in cooling and lubrication.

Similarly, Alsarraf *et al.* [6] observed the impact of nanoparticle on liquid flow properties of Boehmite alumina nanofluid in a horizontal double-channel heat exchanger. The boehmite alumina nanoparticles can be formed in a water / ethylene glycol mixture. In this research, nano fluids and water pass through the annular and tube side of the thermal exchanger, respectively. Besides, they studied the impact of nanoparticle concentrations on the fluid flow properties. Their findings show that the platelet and spherical form lead to the lowest and highest efficiency index of heat exchanger, respectively. Moreover, they found out that when the Reynolds number is equal to 20000, when the concentration of nanoparticles is increased from 0.5 to 2%, the performance of the nanoparticles containing the platelet form and the spherical nanoparticles decrease.

On the other hand, Ranjbarzadeh *et al.* [7] experimentally studied the stability and thermal conductivity of water / fluid silica nanoparticles. Silica nanoparticles were synthesized from the natural eco-friendly rice plant source. In this study, they examined the thermal stability and conductivity of the nanofluid. Some of their results indicate that there is an increase in maximum thermal conductivity of 38.2% at the temperature equal 55°C and at a solid volume fraction of 3%. Moreover, they confirm that this type of nanofluid can be offered as an environmentally friendly alternative for cooling thermal systems.

Dat *et al.* [8] studied the effect of the nanoadditive shape on fluid flow and heat transfer aspects of the γ-AlOOH nanofluid flowing through a sinusoidal wave channel. In this research, they used the two-phase mixture approach to examine. The γ-AlOOH (alumina boehmite) is dispersed in various shapes (such as cylindrical, brick, blade, and platelet) in a 50/50 mixture of water-ethylene glycol as the base liquid. They also studied the effect of the Reynolds number and the concentration volume of the nanoparticles on Nusselt number, where they discovered that the shape of platelets represents the highest performance of heat transfer; they also found that improving the Reynolds number and enhancing the volume fraction of nanoparticles leads to an increase in the number of Nusselt.

Moradikazerouni *et al.* [9] studied the thermal performance of an air-cooled flat heat sink in forced convection using a 3D analytical model. Here, they compare the numerical and experimental results and examine the effective variables for the rate of heat transfer from the surface of the heat sink, where, they also studied the effect of fin height and number of fins on Reynolds number and the heat transfer coefficient under different airflow velocities. Among the most important results they obtained are improving the number and height of fins reduces the surface temperature by 25% and 20%, respectively.

Also, Tian *et al.* [10] investigated the rheological behavior of nanofluid containing CuO / MWCNTs nanoparticles in base liquid water / EG (70:30) at 20-60°C. Nanofluid homogeneous samples are prepared in different nanoparticle volume concentration (0.025, 0.05, 0.1, 0.25, 0.5 and 1%). One of the most prominent results obtained is that the incorporation of nanoparticles into the volume fractions 0.025, 0.05, 0.1, and 0.25 has no impact on the Newtonian behavior of the base liquid, while in the volumetric concentrations 0.5 and 1%, it changes the behavior to non-Newtonian.

Whereas another by Mehdi *et al.* [11] made aims to evaluate the thermo-hydraulic properties of hybrid nanofluids containing nanoparticles of graphene and silver in a heat sink with nanofluids equipped with ribs and secondary channels, their results showed that the use of nanofluids, ribs and secondary channels in the micro-channel, greatly improves the performance of the heat sink. They also confirm that increasing the volume concentration and Reynolds number helps to decrease the temperature and improves the average heat transfer coefficient with convection.

The objective of this paper is to highlight the effectiveness of increasing the surface area of heat exchange between the coolant and the channel walls in the process of cooling electronic components, where fins were added to the channel to increase the contact surface with the nanofluid. In addition, we tested the heat exchange properties of a nanofluid containing copper oxide particles in three mini channels with different exchange surfaces for the purpose of further improving the cooling of the electronic component.

The work presented is organized into five titles: The first title is dedicated to the introduction.

The geometries of the mini channels selected for the test and their dimensioning are represented using ANSYS FLUENT in Title: STUDIED GEOMETRIES.

The third title: MATHEMATICAL FORMULATION presents the governing equations with the boundary conditions. Also, the physical properties of the CuO-H_2O nanofluid and how to calculate them are mentioned.

In the fourth title, the independence of the solution is tested for the simulation, and the results are validated. Also, the simulation results are presented in the form of contours and curves with interpretations.

Finally, a conclusion, which summarizes the main results obtained, is given at the end of this article, with some recommendations for future studies.

STUDIED GEOMETRIES

The geometry of the mini-channel of the cooler is represented by Fig. (**1**). Three different shapes using a fluent software have been studied. The dimensions of the cooler are in the order of 40 x 52mm^2 with a thickness of 6 mm. This mini-channel cooler is composed of 13 channels, the maximum flux of the electronic components is assumed to be constant. The inlet temperature of the CuO-H$_2$O nano-fluid in the three cases of the mini-channels is set at 293K. All outer faces of the cooler are thermally insulated. For reasons of symmetry, only the minichannel has been simulated.

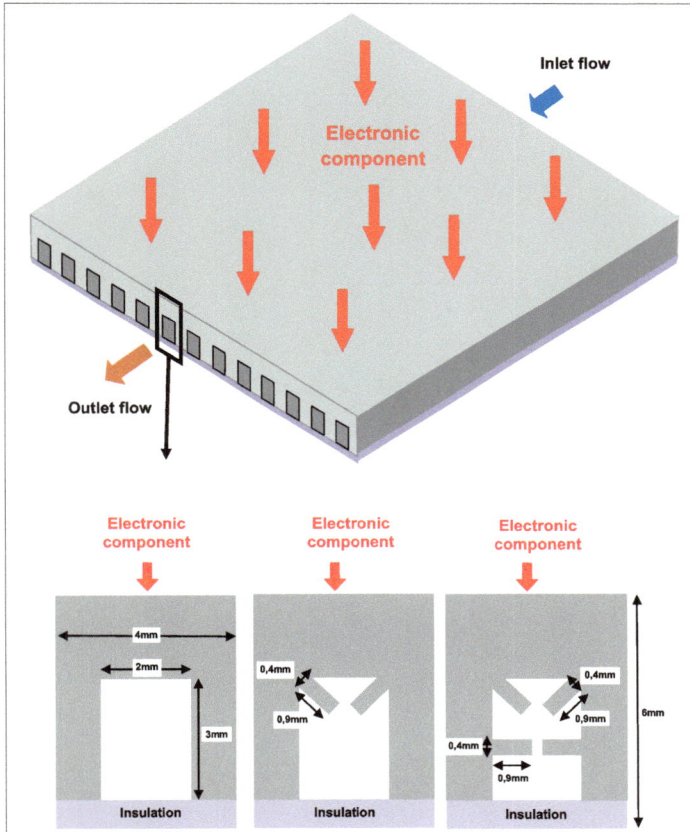

Fig. (1). The different cases of the studied mini channels.

MATHEMATICAL FORMULATION

In this work, we supposed that the flow is stationary. The CuO-water nanofluid is supposed to be Newtonian and incompressible. The radiation heat exchange is negligible. The thermo-physical properties of the CuO-H$_2$O nano-fluid are taken as constant.

The boundary conditions are:

- At the outlet of the mini-channels, the pressure is zéro ($P=0$)
- The velocity components (u, v, w) of the nanofluid at the level of the cannel wall are equal to zero, and no-slip boundary conditions are applied to all mini channel walls.
- The effect of body force and viscosity dissipation is neglected.
- At the inlet flow, the velocity and the temperature of nanofluid are constant.

Governing equations are given in term continuity equation, momentum equation, energy conservation equation and solid equation [12].

The mass conservation is written as follows:

$$\frac{\partial u}{\partial x} + \frac{\partial v}{\partial y} + \frac{\partial w}{\partial z} = 0 \tag{1}$$

where u, v and w are the speed components in x-axis, y-axis, z-axis, respectively.

The momentum equations along the x-axis, the y-axis and the z-axis are written as follows:

x-axis:

$$u\frac{\partial u}{\partial x} + v\frac{\partial u}{\partial y} + w\frac{\partial u}{\partial z} = \frac{1}{\rho_{nf}}\left[-\frac{\partial P}{\partial x} + \mu_{nf}\left(\frac{\partial^2 u}{\partial x^2} + \frac{\partial^2 u}{\partial y^2} + \frac{\partial^2 u}{\partial z^2}\right)\right] \tag{2}$$

y-axis:

$$u\frac{\partial v}{\partial x} + v\frac{\partial v}{\partial y} + w\frac{\partial v}{\partial z} = \frac{1}{\rho_{nf}}\left[-\frac{\partial P}{\partial y} + \mu_{nf}\left(\frac{\partial^2 v}{\partial x^2} + \frac{\partial^2 v}{\partial y^2} + \frac{\partial^2 v}{\partial z^2}\right)\right] \tag{3}$$

z-axis:

$$u\frac{\partial w}{\partial x} + v\frac{\partial w}{\partial y} + w\frac{\partial w}{\partial z} = \frac{1}{\rho_{nf}}\left[-\frac{\partial P}{\partial z} + \mu_{nf}\left(\frac{\partial^2 w}{\partial x^2} + \frac{\partial^2 w}{\partial y^2} + \frac{\partial^2 w}{\partial z^2}\right)\right] \tag{4}$$

where ρ_{nf}, μ_{nf} and P are the density, dynamic viscosity and the pressure of the nano fluid, respectively.

- The energy equation:

$$u\frac{\partial T}{\partial x} + v\frac{\partial T}{\partial y} + w\frac{\partial T}{\partial z} = \alpha_{nf}\left(\frac{\partial^2 T}{\partial x^2} + \frac{\partial^2 T}{\partial y^2} + \frac{\partial^2 T}{\partial z^2}\right) \tag{5}$$

Where T is the temperature of the nano fluid, α_{nf} is the thermal diffusivity of the nano fluid.

- The solid equation

$$\frac{\partial^2 T_S}{\partial x^2} + \frac{\partial^2 T_S}{\partial y^2} + \frac{\partial^2 T_S}{\partial z^2} = 0 \tag{6}$$

The formulas for the calculation of the thermo-physical properties of CuO-Water nano fluid utilized in this study are written as follows:

- The effective thermal conductivity of CuO-Water nano-fluid is approximated as follows [13]:

$$k_{nf} = \frac{k_S + 2k_f - 2\varphi(k_f - k_S)}{k_S + 2k_f + \varphi(k_f - k_S)} k_f \tag{7}$$

Where k_s, k_f and φ are the thermal conductivity of the solid, the thermal conductivity of base fluid (water) and volume fraction, respectively.

- The dynamic viscosity is approximated by Brinkman model as below [13]:

$$\mu_{nf} = \frac{\mu_f}{(1-\varphi)^{2,5}} \tag{8}$$

Where μ_f is the dynamic viscosity of coolant.

>

- The density of CuO-Water nano-fluid is given as [14]:

$$\rho_{nf} = (1 - \varphi)\rho_f + \varphi\rho_S \tag{9}$$

Where ρ_f and ρ_s are the density of the fluid and the solid

- The heat capacitance of CuO-Water nano fluid given as [14]:

$$(\rho C_P)_{nf} = (1 - \varphi)(\rho C_P)_f + \varphi(\rho C_P)_s \tag{10}$$

Where C_p is the specific thermal of the fluid.

The thermo-physical properties of CuO nanoparticles and water are presented in Table **1**.

Table 1. The thermophysical properties of base liquid and CuO nanoparticles at T= 293K [15].

The Thermo-Physical Properties	Water (base fluid)	CuO Nanoparticles
Density (kg/m³)	998.2	6500
Specific heat (Jkg⁻¹ K⁻¹)	4182	535.6
Thermal conductivity (Wm⁻¹K⁻¹)	0.613	20
Dynamic viscosity (kg/m.s)	0.001003	/

RESULTS AND INTERPRETATIONS

The mesh is realized using the fluent software 'Meshing'. After convergence of the calculations of the simulation, the results are represented as follows: Fig. (**2**) shows the evolution of the temperature value of the upper wall of the mini-channel cooler of the second case along the plane of symmetry for the three meshes applied and for Reynolds number equal to 100 and for CuO-water nanofluid with volume concentration equal to 4%. Therefore, it is concluded that the solution is independent of the mesh.

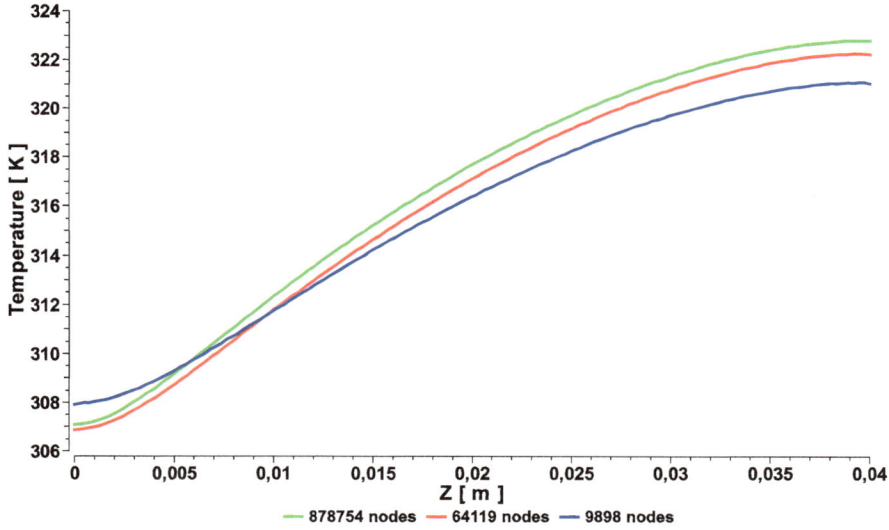

Fig. (2). Influence of the mesh on the temperature of the upper of the minichannel cooler of the 2nd case for Re=100.

Fig. (**3**) shows a comparison of the temperature of this simulation with an experimental study [16]. The temperature increases with the increase of the power dissipated in the chip.

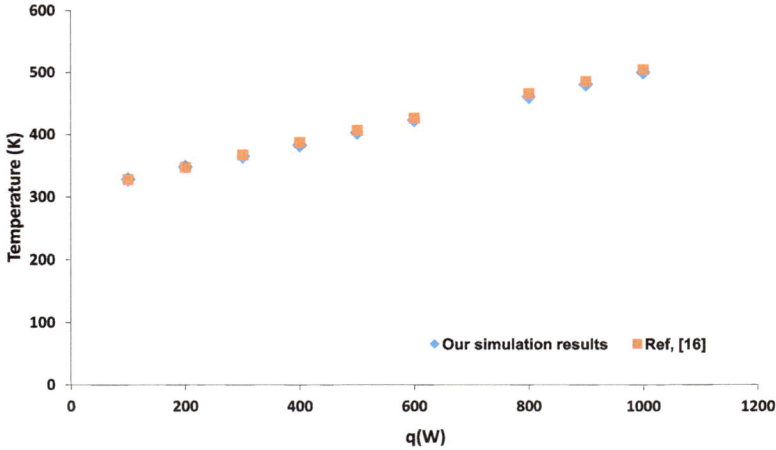

Fig. (3). The temperature according to the power dissipated in the chip.

A good agreement is obtained for the cooler case of copper and water (such as liquid cooling).

Fig. (**4**) shows the evolution of the maximum junction temperature of electronic component as a function of the Reynolds number.

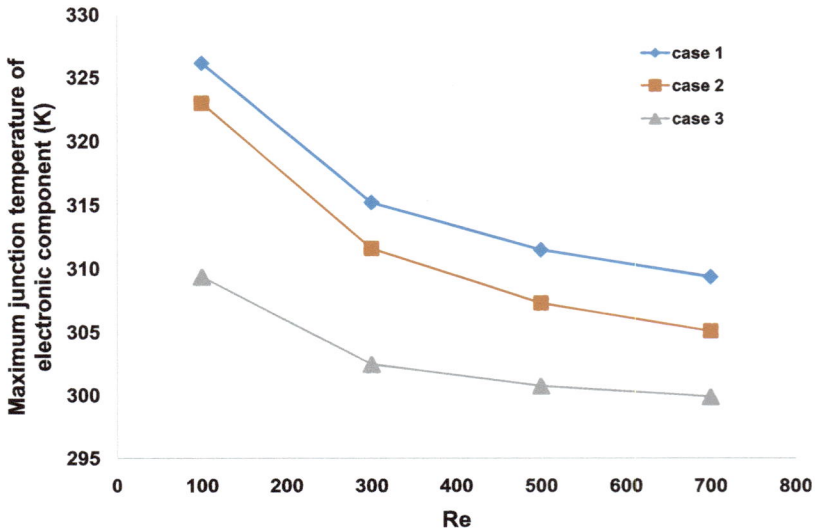

Fig. (4). The values maximum junction temperature of electronic component obtained by simulations for a volume concentration of CuO-water nano-fluid of 4%.

It is observed that the profile of the junction temperature decreases substantially for the three cases of the mini-channels when the Reynolds number increases. We also note that the temperature of the electronic component has the highest values for low values of Re.

Fig. (**5**) shows that for a Reynolds number equal to 100, and the volume concentration of nano-particles equal to 4%, the isotherms become more curved and tighter on the walls of the mini-channel. The comparison between the three cases of minichannels used shows that geometrie of minichannel of third case a greater heat transfer than case 1 and case 2, as shown in Fig. (**6**).

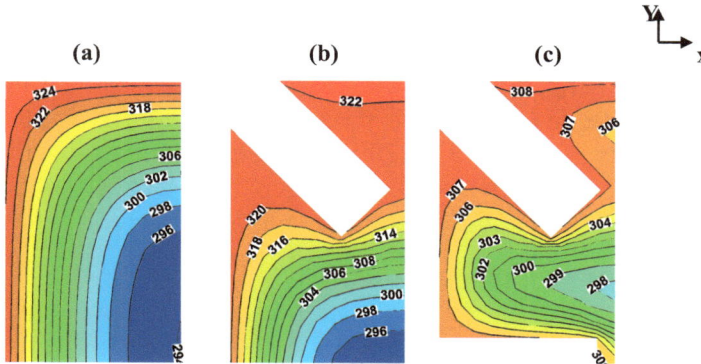

Fig. (5). The distribution of the temperature in outlet of mini channel for three cases studied and for CuO-water nanofluid (Re=100): (a) case1, (b) case 2 and (c) case 3.

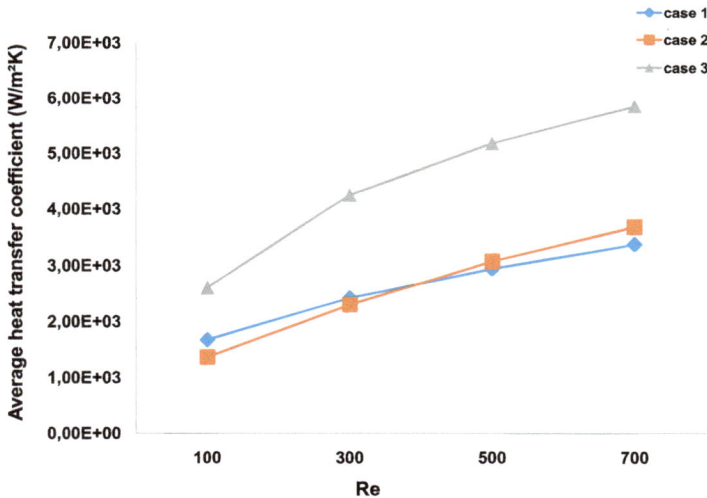

Fig. (6). The average heat transfer coefficients as a function of the Reynolds number for a volume fraction of 4%.

Fig. (**6**) shows the average heat transfer coefficient calculated as a function of Reynolds number for a volume fraction of 4%; the average heat transfer coefficient is proportional to Reynolds number, varying from 100 to 700.

The comparison of the findings of the three examined cases of mini-channels shows that the average heat transfer coefficient of the mini-channel of the third case is greater than those of the mini-channels of the first and second cases.

CONCLUSION

In the present work, the study carried out on the heat exchanges through the different geometries of cross sections of the mini channels of a chiller were studied numerically, using the software Fluent 17.0, depending on the results obtained, we can conclude that for the three cases of mini-channels and with a Reynolds number between 100 and 700, the mini-channels of the third case improve the heat transfer compared to the other cases as well as the maximum temperature value of the junction of the component electronic.

CONSENT FOR PUBLICATION

Not applicable.

CONFLICT OF INTEREST

The author declares no conflict of interest, financial or otherwise.

ACKNOWLEDGEMENTS

Declared none.

REFERENCES

[1]   C.P. Gunnasegaran, H.A. Mohammed, N.H. Shuaib, and R. Saidur, "The effect of geometrical parameters on heat transfer characteristics of micro-channels heat sink with different shapes", *Int. Commun. Heat Mass Transf.,* vol. 37, pp. 1078-1086, 2010.
[http://dx.doi.org/10.1016/j.icheatmasstransfer.2010.06.014]

[2]   H. Jing, L. Zhong, Y. Ni, J. Zhang, S. Liu, and X. Ma, "Design and simulation of a novel high-efficiency cooling heat-sink structure using fluid-thermodynamics", *J. Semicond.,* vol. 36, no. 10, 2015.
[http://dx.doi.org/10.1088/1674-4926/36/10/102006]

[3]   S. Muhammad, and K. Man-Hoe, "Heat transfer enhancement using nano-fluids (Al2O3-H2O) in mini-channel heat sinks", *Int. J. Heat Mass Transf.,* vol. 120, pp. 671-682, 2018.
[http://dx.doi.org/10.1016/j.ijheatmasstransfer.2017.12.075]

[4]   M. Yulin, A. Shahsavar, I. Moradi, S. Rostami, A. Moradikazerouni, H. Yarmand, and N.W.B. Mohd Zulkifli, "The natural convective heat transfer of nano-fluid flow inside an inclined enclosure with conductive walls in the presence of a constant temperature heat source", *Physica A*, p. 123035, 2019.

[5]   A. Asadi, S. Aberoumand, A. Moradikazerouni, F. Pourfattah, G. Żyła, P. Estellé, O. Mahian, S. Wongwises, H.M. Nguyen, and A. Arabkoohsar, "Recent advances in preparation methods and thermophysical properties of oil-based nanofluids: A state-of-the-art review", *Powder Technol.,* vol. 352, pp. 209-226, 2019.
[http://dx.doi.org/10.1016/j.powtec.2019.04.054]

[6]   J. Alsarraf, A. Moradikazerouni, A. Shahsavar, M. Afrand, H. Salehipour, and M. Duc Tran, "Hydrothermal analysis of turbulent boehmite alumina nanofluid flow with different nanoparticle shapes in a minichannel heat exchanger using two-phase mixture model", *Physica A,* vol. 520, pp. 275-288, 2019.
[http://dx.doi.org/10.1016/j.physa.2019.01.021]

[7] R. Ranjbarzadeh, A. Moradikazerouni, R. Bakhtiari, and A.A. Masoud Afrand, "An experimental study on stability and thermal conductivity of water/silica nanofluid: Eco-friendly production of nanoparticles", *J. Clean. Prod.,* pp. 1089-1100, 2019.
[http://dx.doi.org/10.1016/j.jclepro.2018.09.205]

[8] D. Dat, "Vo, Alsarraf J, Moradikazerouni A, Afrand M, Salehipour H, Cong Q. Numerical investigation of γ-AlOOH nano-fluid convection performance in a wavy channel considering various shapes of nanoadditives", *Powder Technol.,* vol. 345, pp. 649-657, 2019.
[http://dx.doi.org/10.1016/j.powtec.2019.01.057]

[9] A. Moradikazerouni, M. Afrand, J. Alsarraf, S. Wongwises, A. Asadi, and T. Khang Nguyen, "Investigation of a computer CPU heat sink under laminar forced convection using a structural stability method", *Int. J. Heat Mass Transf.,* vol. 134, pp. 1218-1226, 2019.
[http://dx.doi.org/10.1016/j.ijheatmasstransfer.2019.02.029]

[10] Z. Tian, S. Rostami, R. Taherialekouhi, A. Karimipour, A. Moradikazerouni, H. Yarmand, and W.B. Nurin Mohd Zulkifli, "Prediction of rheological behavior of a new hybrid nanofluid consists of copper oxide and multi wall carbon nanotubes suspended in a mixture of water and ethylene glycol using curve-fitting on experimental data", *Physica A,* vol. 549, p. 124101, 2020.
[http://dx.doi.org/10.1016/j.physa.2019.124101]

[11] B. Mehdi, J. Mohammad, and G. Marjan, "Efficacy of a hybrid nanofluid in a new microchannel heat sink equipped with both secondary channels and ribs", *J. Mol. Liq.,* vol. 273, pp. 88-98, 2019.
[http://dx.doi.org/10.1016/j.molliq.2018.10.003]

[12] M.K. Moraveji, R.M. Ardehali, and A. Ijam, "CFD investigation of nanofluid effects (cooling performance and pressure drop) in mini-channel heat sink", *Int. Commun. Heat Mass Transf.,* vol. 40, pp. 58-66, 2013.
[http://dx.doi.org/10.1016/j.icheatmasstransfer.2012.10.021]

[13] S. Fohanno, G. Polidori, and C. Popa, *Nanofluides et transfert de chaleur par convection naturelle.* Université de reims champagne-Ardenne: France, 2012.

[14] K. Khanafer, K. Vafai, and M. Lightstone, "Buoyancy-driven heat transfer enhancement in a two dimensional enclosure utilizing nanofluids", *Int. J. Heat Mass Transf.,* vol. 46, pp. 3639-3653, 2003.
[http://dx.doi.org/10.1016/S0017-9310(03)00156-X]

[15] H.A. Mohammed, P. Gunnasegaran, and N.H. Shuaib, "The impact of various nanofluid types on triangular microchannels heat sink cooling performance", *Int. Commun. Heat Mass Transf.,* vol. 38, pp. 767-773, 2011.
[http://dx.doi.org/10.1016/j.icheatmasstransfer.2011.03.024]

[16] A. Yvan, K-L. Afef, C. David, D. Emanuelle, C. Jean, and M. Tawk, *Study of a system of cooling of electronic components of power by liquid metal.* University of Pierre and Marie Curie: Paris, VI, 2010.

CHAPTER 3

Influence of the Force Delivery of Orthodontic NiTi Arch Wire on its Tribological Behavior

Ines Ben Naceur[1,*] and **Khaled Elleuch**[1]

[1] *National Engineering School of Sfax, Department of Materials Engineering and Environment (LGME) ENIS, B.P.W.1173 Sfax, Tunisia*

Abstract: The complexity of orthodontic treatments requires archwires with specific biomechanical properties according to the different stages of therapy. Thanks to their wide elastic zone and low stiffness, superelastic NiTi alloy is used in the leveling and alignment phases. The friction that accompanies the beginning of treatment is a very complicated phenomenon, since in the presence of arch misalignments, the present normal force, which compresses the orthodontic archwire-bracket couple, is very dependent on the clinical situation. This study aims to identify the friction responses and the degradation mechanisms of a superelastic NiTi orthodontic archwire, as a function of the applied normal load. The latter represents the charges delivered by the archwire during its unloading, all through the first phases of treatment. Circular and rectangular samples with the most commonly used dimensions have been tested in a dry environment at room temperature. The results showed that the wear of the NiTi alloy was amplified as a function of the normal force applied for the two tested archwire shapes. Indeed, the degradation regimes observed by scanning electron microscopy present a transition, by increasing the load from a mainly adhesive regime to a more complex situation, in which wear by adhesion is accompanied by abrasive and delamination wear.

Keywords: Abrasion, Adhesion, Arch wire, Delamination, Friction, NiTi alloy, Normal Load, Orthodontic, Scanning electron microscopy, Wear.

INTRODUCTION

The exploitation of the superelasticity of shape memory alloys for the correction of dental malocclusions in orthodontics therapy is a very successful application, since it allows obtaining an optimal tooth movement by strictly controlling and drastically reducing the treatment [1]. Thanks to their low modulus of elasticity,

---

* **Corresponding author Ines Ben Naceur:** National Engineering School of Sfax, Department of Materials Engineering and Environment (LGME) ENIS, B.P.W.1173 Sfax, Tunisia; E-mail:inesbennaceurtounsi@gmail.com

Zied Driss (Ed.)

high resilience and elastic limit, dental archwires made of superelastic Nickel-Titanium (NiTi) alloy are generally used in the first phases of orthodontic treatment [2]. During the alignment and/or levelling phase of treatment, the archwire and brackets are subjected to the frictional forces opposed to the movement of teeth correction due to various oral functions such as chewing. The generated amount of this friction is proportional to the applied normal force which compresses the two surfaces in contact [3]. In this case, the normal forces responsible for correcting malpositioned teeth correspond mainly to the forces induced by the unloading of the NiTi archwire, through the slot of the bracket, following its deflection [4 - 6]. To ensure the desirable movement of the teeth, the resultant of these forces must overcome the friction generated at the interface [7]. Drescher *et al.* [8] have shown that as the reduction in wire deactivation force can reach 50% due to the friction, therefore tooth movement could be impeded. Moreover, this phenomenon is even frustrating because friction will induce the wear of the brackets and arch-wire materials which mainly consist of NiTi alloy and stainless steel. This wear favors the spread of toxic metals like nickel through the human body [9].

The currently available information pertaining to the friction of NiTi-wire stainless steel brackets combinations has been derived primarily from displacing a brack*et al*ong a guiding wire or pulling a wire through a bracket-slot using a standardized universal testing machine [10, 11]. This technique is based on a unidirectional single linear sliding motion between brackets and wire, which does not reflect the real situation. Various oral functions as chewing, swallowing, speaking, *etc.*, result in periodic and repetitive minute-motion at the bracket/archwire interfaces, several thousand times each day [12]. In recent years, few published papers have focused on the tribological properties of this couple using a pin on disk type friction tester [13, 14]. However, these studies did not consider the leveling phase of the treatment and therefore, they did not assess the effect of the normal force delivered by the NiTi archwire, for different deflection levels, on its friction and wear behavior.To overcome some of these shortcomings, a rotative Tribometer was adapted to investigate the tribological behavior of the NiTi alloy sliding against a flat 316 stainless steel piece simulating an orthodontic bracket.

The aim was to compare, in dry conditions and at room temperature, the effect of the normal force delivered by the unloading of the NiTi archwire during the leveling phase of treatmet, on its tribological behavior using a rotating tribometer.

MATERIALS AND METHODS

In this study, we used rectangular (0.46×10^{-3}m * 0.64×10^{-3}m) and circular (0.46×10^{-3}m) superelastic NiTi archwires with an austenitic structure at room temperature. The tribological behavior of the tribo-contact was analyzed using an adapted rotating tribometer. In fact, a model matching the shape of a straight archwire was designed. A small slot has been machined at the concave part in order to embed archwire, and therefore to prevent it from slipping during the tests. This model was mounted in the fixed part of the tribometer, while the brackets replaced by a flat 316 stainless steel parts are fixed by screws on the rotating disc. An overview of the tribometer is shown in Fig. (**1**). A linear velocity of 68×10^{-3} m/s is imposed with a track radius of 3×10^{-3}m for 12 000 cycles. The tests were carried out in dry condition and at room temperature. For each set of conditions, 3 tests were conducted. The loads delivered for the circular and rectangular archwires for three deflection levels, of the leveling phase, are estimated in a previous study [6] by finite elements method using the Abaqus software and summarized in Table **1**.

Table 1. Unloading normal force of a deflected orthodontic archwire at 25°C [6].

Deflection ($*10^{-3}$ m)	Unloading Normal Force (N)	
	Rectangular Arch Wires	Circular Arch Wires
0.5	2.25	0.6
1.5	3.4	1
2.5	9.4	2.2

The micro-hardness tests were carried out using a "Wolpert Wilson 402 MVD" type micro hardness tester. The measurements were carried out on NiTi samples exhibiting a completely austenitic initial state at room temperature with an applied load of 0.5 Kg. Due to the small dimensions of the orthodontic wires, they were coated with a mixture containing hardener and resin in sufficient quantity to have a fluid mixture. A measurement of the mass loss was carried out using a precision of 10^{-4} balance. At the end of each test, the sample is cleaned with ethanol in an ultrasonic bath to remove non-adherent wear particles. The wear mechanisms will be analyzed based on the observations of the wear tracks of NiTi samples using a scanning electron microscope (JEOL JSM 5400). Before the friction tests, the stainless steel flat pieces, having an average hardness of 160 HV, are subjected to finishing polishing operations by diamond paste felt papers. Its chemical composition was determined using a spectrometer (Jobin Yvon JY 48). Table **2** lists the content of the main elements in percentage by mass.

Table 2. Chemical composition: AISI 316.

C (%)	Si (%)	Mn (%)	P (%)	S (%)	Cr (%)	Ni(%)	Mo (%)
0.084	0.54	1.8	0.018	0.0038	17.28	10.51	2.02

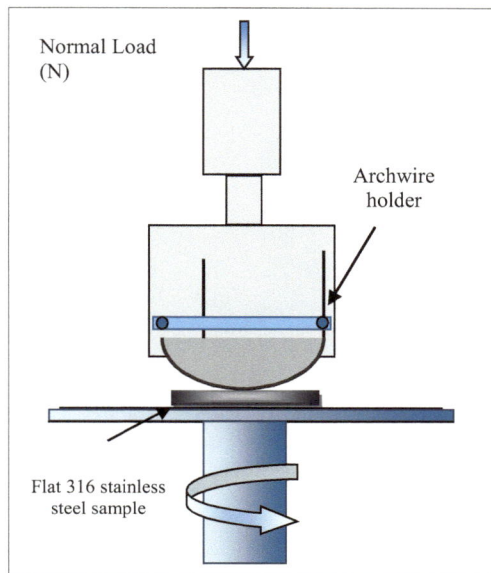

Fig. (1). Configuration used for sliding experiments: rotating wear tests.

RESULTS

Microhardness Test

From the obtained results for different measuring positions Fig. (**2**), it can be noted that the material is homogeneous, whose micro-hardness recorded at the surface of the sample is around 350 HV.

Tribological Response of NiTi Archwires

This section describes the effect of increasing normal load on the friction and wear behavior of NiTi alloy sliding against a 316 stainless steel flat sample. The applied loads represent the actual forces magnitude that receives a tooth during the unloading of the NiTi archwire, for each corresponding deflection, during the first stages of orthodontic treatment (Table **1**). These corresponding values were added to the force induced by the elastic ligatures which compress the orthodontic arch-bracket couple which is estimated by several researchers to 2 N [15].The study of the effect of vertical deflection on the results was presented in a previous study carried out by Bennaceur *et al*. [16]. Figs. (**3a** - **3b**) show the effect of increasing the normal load on the average friction coefficient of the tested tribo-contact, respectively, for circular and rectangular NiTi archwires.

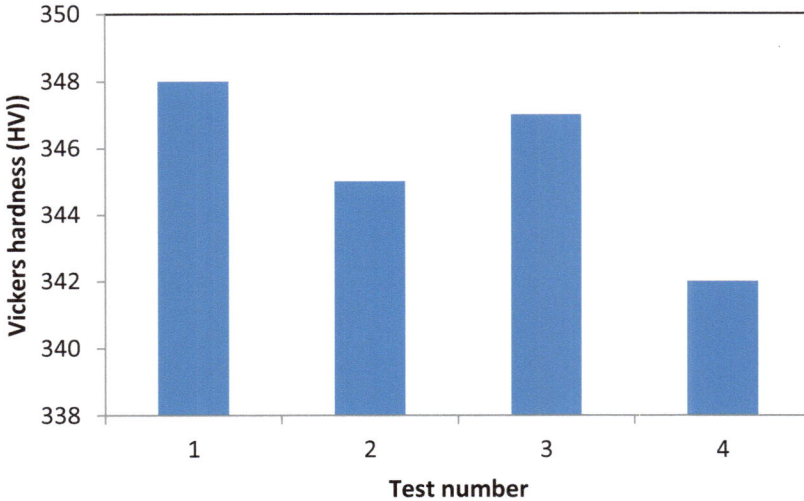

Fig. (2). Vickers hardness of the tested NiTi alloy.

The different curves of this figure show similar patterns. Indeed, the overall changes in the coefficient of friction over time for the two tested archwires shapes are therefore very similar. Two friction phases are observed, the first of which is characterized by a drop in friction after a highly disturbed transient stage, and the second is rather stable, reaching an almost stationary value of the friction coefficient. Such typical behavior could be related to changes that emanate from the progression of damage to the interfaces of the two friction surfaces during sliding. Actually, during the first sliding cycles, *i.e.*, the running-in phase, the surfaces are therefore accommodated by leveling the asperities and the roughness. Thus, the increase in the coefficient of friction can be attributed to an initial abrasive contact between the wire and the bracket, or to the energy dissipated by friction necessary for the wear debris formation. The friction coefficient then remains stable until the end of the test [16, 17]. It is trusty to note that the coefficient of friction fluctuates significantly. The origins of the fluctuations are not identified with certainty. They could be explained either by the instability of the two bodies in contact (stick-slip phenomena), or by the modification of the topography of the third body at the contact interface [18].

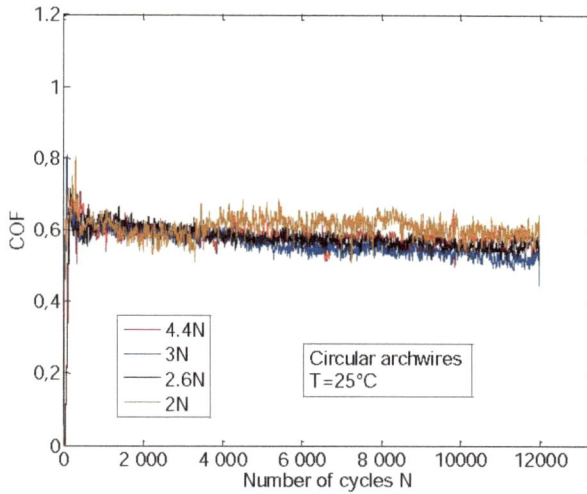

(a)

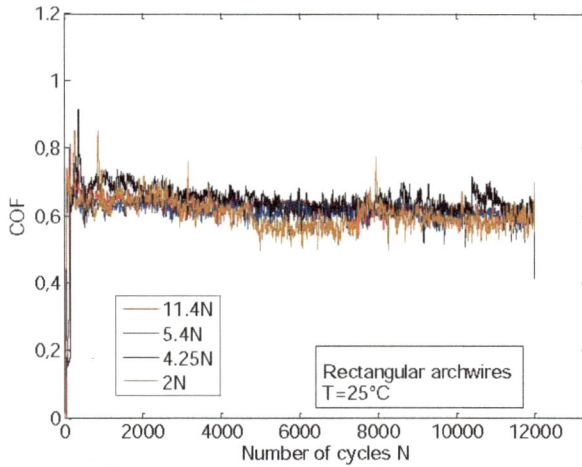

(b)

Fig. (3). Coefficient of friction with respect to sliding cycles for different normal loads, (a) circular and (b) rectangular archwires, 12.000 sliding cycles.

Fig. **(4)** displays the variation of the stabilized average friction coefficient, carried out in dry condition and at room temperature, as a function of the applied normal load. The friction coefficient of the orthodontic tested couple remains around the value of 0.6 for the rectangular and circular wires, regardless of the applied normal force Fig. (**4**). The values of the coefficient of friction are influenced by the change in the surface condition during sliding. Besides, since the dry friction of the tested tribo-contact is very severe, it is possible that the phenomenon of leveling asperities and refinement of roughness occurs rapidly and in the same way at low and high load, and therefore the friction coefficient remains almost constant for all applied loads.

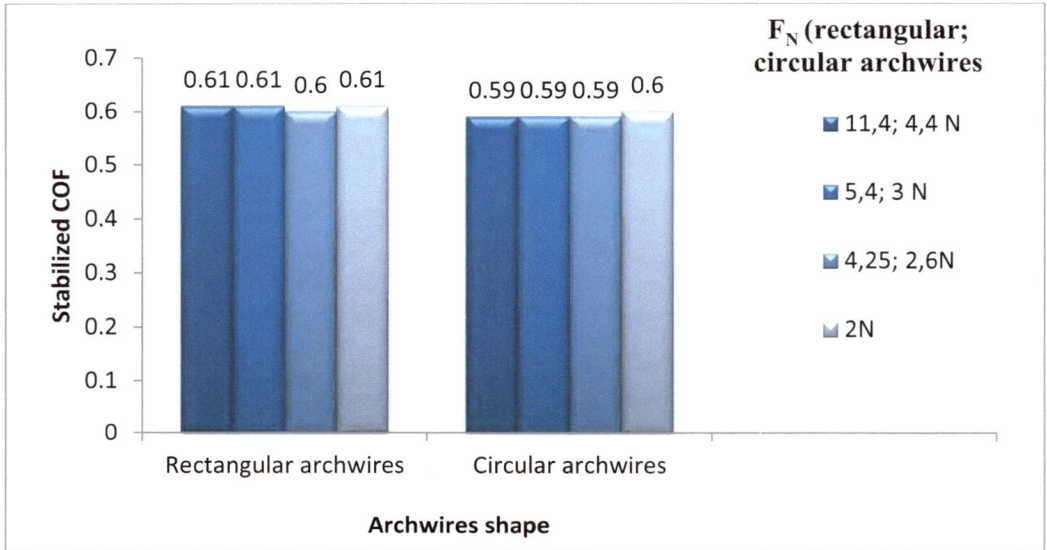

Fig. (4). Stabilized friction coefficient as a function of applied normal load for rectangular and circular archwires.

Fig. (**5**) shows the effect of the applied normal force on the weight loss of circular Fig. (**5a**) and rectangular Fig. (**5b**) NiTi archwires after 12.000 friction cycles. The obtained results showed that the wear resistance of rectangular archwires was better than that of the circular ones. In fact, the corresponding mass losses at the applied normal load of 2N are 0.07 10^{-6} kg and 0.135 10^{-6} kg, respectively. It is noteworthy to mention that while the circular wire initially has a contact point, the rectangular one has a line contact. Hence, the resultant of the normal force is more intense in the case of point contact. This force could potentially indent the circular archwire, and can thus significantly reduce its resistance to wear.

On the other hand, it is to be noted that the wear of the NiTi alloy increases with the increase in the normal load. Indeed, under 5.4 and 11.4 N applied load, the weight loss of the NiTi rectangular archwires increases with 50% and 80%, respectively, compared with the test carried out at 4.25N.

These results can be explained by the increase in the real contact area with the increase of the applied normal force. This finding obeys to Archard's law [19], stipulating that R = K (P / H) v: where R stands for wear rate (volume/time), K for wear coefficient, P for pressure, H for hardness, and v for velocity. According to this law, the wear speed is proportional to the applied normal force.

This result allows us to consider that the greater the applied load is, the more the wear mechanisms present are amplified, which is probably the result of the generation of high concentration of compressive stress in the contact zone, thus promoting the delamination mechanism. In fact, these compressive stresses produce the rupture of cracks in the maximum shear zones. Furthermore, a material surface necessarily consists of asperities. Normal and tangential loads are transmitted through the interacting asperities when two surfaces come into contact. It has been demonstrated [20] that during a sliding wear process, cracks nucleate and then propagate, leading to an eventual material removal from the surface. As explained by Ashby and Jones [20], this phenomenon is enhanced with the increase in the normal load.

The values of the coefficient of friction as well as the wear rate of the NiTi alloy are influenced by the change in the state of the rubbing surfaces. Thus, it is appropriate to explore the wear mechanisms by the analysis of the wear tracks in order to find an explanation for the found trends. We therefore conducted a scanning electron microscopy examination of the wear trace following 12.000 sliding cycles. The obtained results are represented for the circular and rectangular archwires in Figs **6** and **7**, respectively. A higher magnification of the wear trace of the circular NiTi archwire, tested with an applied load of 2N (Fig. **6a**), reveals the presence of many black spots along the sliding direction. The obtained results lead to consider the formation of a transfer layer (darker regions) from 316 stainless steel on the NiTi alloy as the first stage of degradation. A large number of triboparticles detached from the stainless steel sample, having a lower hardness than the tested NiTi alloy (160 Hv against 350 Hv for the NiTi alloy), subsequently adhere to the NiTi archwire. The detachment of the first debris does not result in the formation of wear particles, but rather participates in the formation of a new surface layer. These findings represent a piece of evidence for adhesive wear, which is usually accompanied by the transfer of metal from one surface to another, generally from the softer material to the harder material [21, 22], following a local rupture on a weak point of one of the substrates. This adhesive wear is observed mainly at low loads. These findings accord well with those obtained by Liu *et al.* [23]. Indeed, they found that the worn surfaces of the NiTi alloy sliding against the 304 stainless steel alloy also exhibit the characteristics of adhesive wear with the development of scratches parallel to the direction of sliding. They have analyzed the worn surface of the tested NiTi alloy using the energy dispersive X-ray spectroscopy. Their results have demonstrated that a considerable quantity of Iron and Chromium was transported to the surface of the NiTi alloy during sliding. The findings pertaining to the formation of this superficial layer transferred onto the worn surface of the NiTi alloy, from the stainless steel surface, was also recorded by Abedini *et al.* [22], and Grosgogeat *et al.* [15].

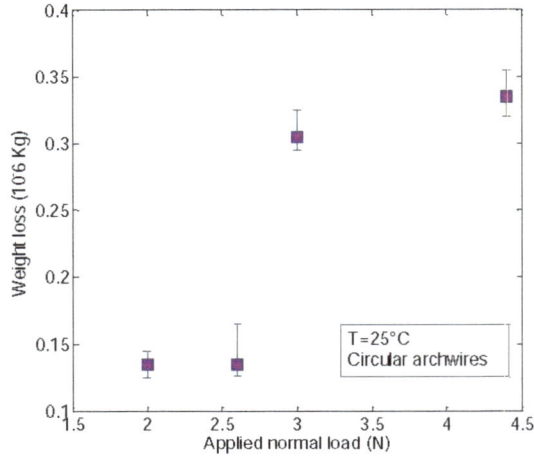

(a)

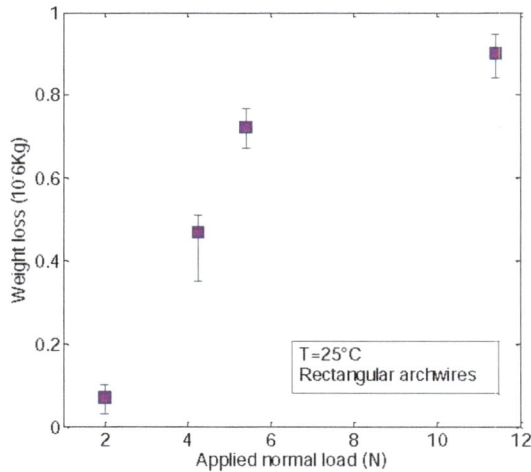

(b)

Fig. (5). Evolution of the weight loss of (a) circular NiTi and (b) rectangular arch wires as a function of the normal applied load after 12.000 sliding cycles.

Therefore, severe abrasive wear and plastic deformation appear as a result of repetitive friction cycles on the wear trace. This is clear from the SEM observations of all the tested combinations (Figs **6** - **7**). This abrasive wear is manifested by the appearance of scratches and grooves on the friction facies, which are shallow (not much profound) at low applied loads for the circular and rectangular wires and much more obvious and accentuated for extreme test conditions (Figs **6b** - **7b**). Indeed, the SEM analysis of the wear tracks on the circular and rectangular NiTi archwires, tested with a load of 4.4N and 11N (Figs **6a** - **7a**), respectively, reveal a plastic deformation of the material as well as the presence of abundant deep grooves on the wear track oriented along the sliding direction. These grooves are therefore the result of a plowing action by the counterface and / or the detachment of wear debris as a third body [24]. The increase in the normal force appears to change the wear mechanism of the NiTi alloy towards predominantly abrasive wear [25].

In Alfonso *et al.* [26]'s study of the wear of NiTi orthodontic wires sliding in saliva at 37°C, abrasion wear by the action of the third body was similarly found. Furthermore, Abedini *et al.* [22] have shown that plowing and abrasion wear mechanisms could be important in the sliding wear of the NiTi alloy and these damages are more accentuated with the increase of the applied load. In the same vein, Lin and his colleagues [27] have demonstrated that the friction facies of the NiTi alloy is characterized by severe damage of the wear track with the production of a large number of triboparticles in an abrasive adhesive wear mechanism. Furthermore, they have found that these damages are more accentuated by increasing the applied load [22, 27].

Through a higher magnification at the end of the wear mark (Fig. **6a**), it is possible to discern the triboparticles detached and released outside the friction track. Hence, it appears that the material inside the grooves has been moved to the edges forming a topography that can be described as extruded edges. Moreover, the signs of wear by delamination and chipping can be observed at low and high loads. The examination of the corresponding rubbed surfaces shows that this mechanism of particle detachment, augmenting especially with the increase in the normal force (Figs **6b** - **7b**), gives rise to elementary particles of different sizes ranging from the micrometric scale to the nanometric scale. This is clear on the magnification carried out on the wear track of the circular wire represented in Fig. (**6b**). Actually, the detachment of the wear particles like sheet was generated, as shown in Fig. (**6a**), indicating that the lamellae of NiTi are completely detached from the surface due to the delamination wear.

The friction between two surfaces can be considered as a cyclic phenomenon given the repetition of stresses produced by sliding cycles, thus inducing the phenomenon of delamination. Besides, the wear occurs by the deformation of the surface layer, nucleation of cracks at the level of the subsurface and its propagation parallel to the sliding direction. Moreover, the degraded surface reveals scale-shaped areas from which the microplates of material have detached. Subsequently, under the friction action, the flat plates fragment leads to the complete detachment of third body debris, which constitutes a secondary source flow and entertains, in turn, the tribological circuit [22, 27].

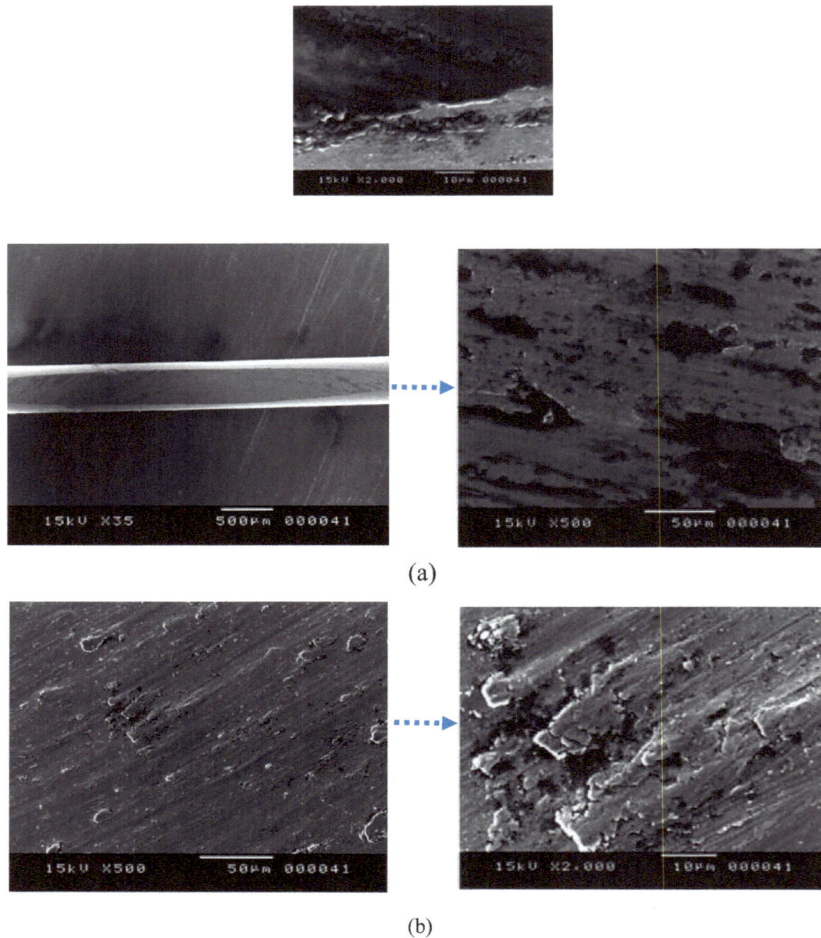

(a)

(b)

Fig. (6). SEM micrographs of the worn surfaces of circular NiTi arch wires tested at 25°C as a function of applied normal load: (a) 2N and (b) 4.4N, 12.000 cycles.

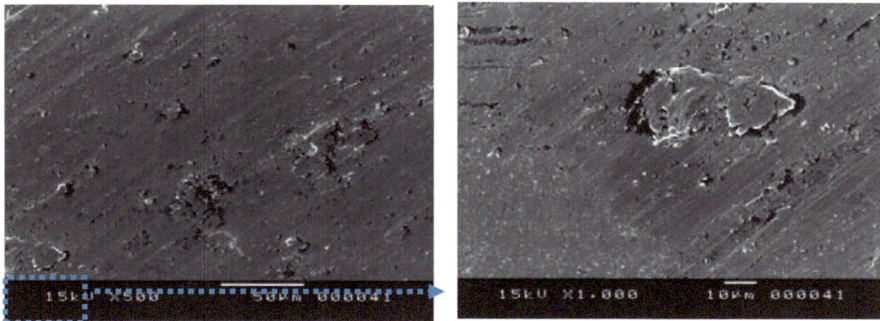

(a)

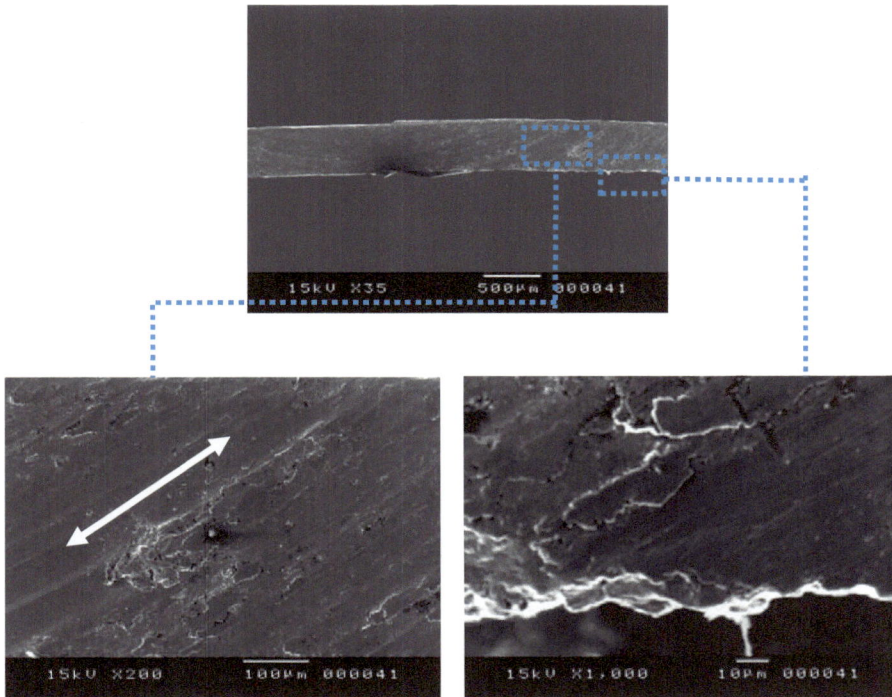

(b)

Fig. (7). SEM micrographs of the worn surfaces of rectangular NiTi arch wires tested at 25°C as a function of applied normal load: (a) 2N and (b) 11N, 12.000 cycles.

The characterization of the worn surfaces of superelastic NiTi alloy has been studied by Farhat *et al.* [28]. Concerning the delamination wear, it is identified as the dominant wear mechanism in dry wear test. According to these authors, the delamination of the NiTi alloy is promoted by the mismatch in the elastic properties across the cross section of the alloy. The upper layer undergoes a plastic deformation and loses its superelastic property. However, subterranean layers still have superelastic behavior, *i.e.*, they still have a large elastic covering. This difference in mechanical properties leads to the formation of cracks in the subsurface which, when propagated, form wear debris by delamination [28]. A previous study [22, 29] has also shown the predominance of delamination wear for NiTi alloys with the increase in the normal load.

CONCLUDING REMARKS

In this study, the tribological behavior of NiTi superelastic orthodontic archwires sliding against a 316 stainless steel plate, was investigated as a function of the normal load. These applied forces represent the forces delivered by the NiTi archwires during its unloading all along the orthodontic leveling phase. The tests were conducted in dry conditions, at room temperature and for different archwires shapes. The examinations of the wear tracks provided the necessary information to understand the wear mechanisms of the tested tribo-contact. Three wear mechanisms have been identified as the dominant mechanisms in the sliding wear of the NiTi alloy, namely the adhesion, abrasion and delamination. In a first stage, the wear appears to be of an adhesive nature. Detached particles from repetitive sliding cycles give rise to a severe abrasive wear. Furthermore, the damage of the wear trace was found to be much more severe by increasing the applied normal load. This probably results in the accumulation of this abrasive effect, and thus leading to an increase in the wear rate of the NiTi alloy.

CONSENT FOR PUBLICATION

Not applicable.

CONFLICT OF INTEREST

The author declares no conflict of interest, financial or otherwise.

ACKNOWLEDGEMENTS

Declared none.

REFERENCES

[1]   F. Auricchio, and E. Sacco, "A temperature-dependent beam for shape-memory alloys: constitutive modeling, finite-element implementation and numerical simulations", *Comput. Methods Appl. Mech. Eng.*, vol. 174, pp. 171-190, 1999.

[http://dx.doi.org/10.1016/S0045-7825(98)00285-0]

[2]   B. Coluzzi, A. Biscarini, and L.D. Massoa, "Phase transition features of NiTi orthodontic wires subjected to constant bending strains", *J. Alloys Compd.,* vol. 233, pp. 197-05, 1995.
[http://dx.doi.org/10.1016/0925-8388(95)01975-8]

[3]   P.F. Friction, *Sci. Am.,* vol. 184, pp. 54-60, 1951.

[4]   S. Dechkunakorn, N. Viriyakosol, and N. Anuwongnukroh, "Residual force of orthodontic elastomeric ligature", *Adv. Mat. Res.,* vol. 378, pp. 674-680, 2012.

[5]   I. Bennaceur, and K. Elleuch, "Tribological properties of deflected NiTi superelastic archwire using a new experimental set-up: Stress-induced martensitic transformation effect", *Tribol. Int.,* vol. 146, 2020.: 10603.
[http://dx.doi.org/10.1016/j.triboint.2019.106033]

[6]   I.B. Naceur, A. Charfi, T. Bouraoui, and K. Elleuch, "Finite element modeling of superelastic nickel-titanium orthodontic wires", *J. Biomech.,* vol. 47, no. 15, pp. 3630-3638, 2014.
[http://dx.doi.org/10.1016/j.jbiomech.2014.10.007] [PMID: 25458153]

[7]   G.F. Andreasen, and T.B. Hilleman, "An evaluation of 55 cobalt substituted Nitinol wire for use in orthodontics", *J. Am. Dent. Assoc.,* vol. 82, no. 6, pp. 1373-1375, 1971.
[http://dx.doi.org/10.14219/jada.archive.1971.0209] [PMID: 5280052]

[8]   D. Drescher, C. Bourauel, and H.A. Schumacher, "Frictional forces between bracket and arch wire", *Am. J. Orthod. Dentofacial Orthop.,* vol. 96, no. 5, pp. 397-404, 1989.
[http://dx.doi.org/10.1016/0889-5406(89)90324-7] [PMID: 2816839]

[9]   A. David, and D. Lobner, "In vitro cytotoxicity of orthodontic archwires in cortical cell cultures", *Eur. J. Orthod.,* vol. 26, no. 4, pp. 421-426, 2004.
[http://dx.doi.org/10.1093/ejo/26.4.421] [PMID: 15366387]

[10] U.H. Doshi, and W.A. Bhad-Patil, "Static frictional force and surface roughness of various bracket and wire combinations", *Am. J. Orthod. Dentofacial Orthop.,* vol. 139, no. 1, pp. 74-79, 2011.
[http://dx.doi.org/10.1016/j.ajodo.2009.02.031] [PMID: 21195280]

[11] R.H. Higa, N.T. Semenara, J.F.C. Henriques, G. Janson, R. Sathler, and T.M. Fernandes, "Evaluation of force released by deflection of orthodontic wires in conventional and self-ligating brackets", *Dental Press J. Orthod.,* vol. 21, no. 6, pp. 91-97, 2016.
[http://dx.doi.org/10.1590/2177-6709.21.6.091-097.oar] [PMID: 28125144]

[12] S. Braun, M. Bluestein, B.K. Moore, and G. Benson, "Friction in perspective", *Am. J. Orthod. Dentofacial Orthop.,* vol. 115, no. 6, pp. 619-627, 1999.
[http://dx.doi.org/10.1016/S0889-5406(99)70286-6] [PMID: 10358243]

[13] C. Rapiejko, S. Fouvry, B. Grosgogeat, and B. Wendler, "A representative ex-situ fretting wear investigation of orthodontic arch-wire/bracket contacts", *Wear,* vol. 266, pp. 850-857, 2009.
[http://dx.doi.org/10.1016/j.wear.2008.12.013]

[14] T. Kang, S.Y. Huang, J.J. Huang, Q.H. Li, D.F. Diao, and Y.Z. Duan, "The effects of diamond-like carbon films on fretting wear behavior of orthodontic archwire-bracket contacts", *J. Nanosci. Nanotechnol.,* vol. 15, no. 6, pp. 4641-4647, 2015.
[http://dx.doi.org/10.1166/jnn.2015.9788] [PMID: 26369091]

[15] B. Grosgogeat, E. Jablonska, and J.M. Vernet, "Tribological response of sterilized and un-sterilized orthodontic wires", *Mater. Sci. Eng. C,* vol. 26, pp. 267-272, 2006.
[http://dx.doi.org/10.1016/j.msec.2005.10.050]

[16] I. Bennaceur, and K. Elleuch, "Effects of saliva addition on the wear resistance of deflected NiTi archwire for biomedical application", *Mater. Lett.,* vol. 268, 2020.127550
[http://dx.doi.org/10.1016/j.matlet.2020.127550]

[17] T. Kagnaya, C. Boher, and L. Lambert, "Wear mechanisms of WC–Co cutting tools from high-speed

tribological tests", *Wear,* vol. 267, pp. 890-897, 2009.
[http://dx.doi.org/10.1016/j.wear.2008.12.035]

[18]   Y. Berthier, *Wear, Materials, Mechanisms and Practice.* John Wiley & Sons, 2005.

[19]   J.F. Archard, "Contact and rubbing of flat surfaces", *J. Appl. Phys.,* vol. 24, pp. 981-988, 1953.
[http://dx.doi.org/10.1063/1.1721448]

[20]   MF Ashby, and DRH Jones, " Engineering Materials 1: An Introduction to Their Properties and
Applications: Bodmin Cornwall",

[21]   K.L. Johnson, The mechanics of adhesion, deformation and contamination in friction.*Dissipative
Processes in Tribology.,* D. Dowson, C.M. Taylor, T.H.C. Childs, M. Godett, G. Dalmaz, Eds., , 1994,
pp. 21-33.
[http://dx.doi.org/10.1016/S0167-8922(08)70294-3]

[22]   M. Abedini, H.M. Ghasemi, and N.M. Ahmadabadi, "Effect of normal load and sliding distance on the
wear behavior of NiTi alloy", *Tribol T,* vol. 55, pp. 677-684, 2012.
[http://dx.doi.org/10.1080/10402004.2012.688166]

[23]   R. Liu, and D.Y. Li, "Experimental studies on tribological properties of pseudoelastic TiNi alloy with
comparison to stainless steel 304", *Metall Materi Trans A,* vol. 31, pp. 2773-2783, 2000.
[http://dx.doi.org/10.1007/BF02830337]

[24]   G. Spinler, Ed., *Conception des machines, Principes et applications, 1 statique.* vol. Vol. 1. Epfl Press,
2002, pp. 35-37.

[25]   S. Gialanella, G. Ischia, and G. Straffelini, "Phase composition and wear behavior of NiTi alloys", *J.
Mater. Sci.,* vol. 43, pp. 1701-1710, 2008.
[http://dx.doi.org/10.1007/s10853-007-2358-3]

[26]   M.V. Alfonso, E. Espinar, J.M. Llamas, E. Rupérez, J.M. Manero, J.M. Barrera, E. Solano, and F.J.
Gil, "Friction coefficients and wear rates of different orthodontic archwires in artificial saliva", *J.
Mater. Sci. Mater. Med.,* vol. 24, no. 5, pp. 1327-1332, 2013.
[http://dx.doi.org/10.1007/s10856-013-4887-4] [PMID: 23440428]

[27]   H.C. Lin, J.L. He, and K.C. Chen, "at al. Wear characteristics of TiNi shape memory alloys", *Metall
Materi Trans A,* vol. 28, pp. 1871-1877, 1997.
[http://dx.doi.org/10.1007/s11661-997-0117-3]

[28]   Z.N. Farhat, and C. Zhang, "The Role of reversible martensitic transformation in the wear process of
TiNi shape memory alloy", *Tribol T,* vol. 53, pp. 917-926, 2010.
[http://dx.doi.org/10.1080/10402004.2010.510620]

[29]   C. Zhang, and Z.N. Farhat, "Sliding wear of superelastic TiNi alloy", *Wear,* vol. 267, pp. 394-00,
2009.
[http://dx.doi.org/10.1016/j.wear.2008.12.093]

CHAPTER 4

Reynolds Number Effects on the Flow through a Savonius Wind Rotor

Sobhi Frikha[1,*], **Mariem Lajnef**[1] and **Zied Driss**[1]

[1] *Laboratory of ElectroMechanical Systems (LASEM), National Engineering School of Sfax (ENIS), University of Sfax, 3038 Sfax, Tunisia*

Abstract: In this article, we investigate the influence of the Reynolds number on the flow around a Savonius wind rotor. In particular, we have studied various regimes defined by different Reynolds numbers. Four different Reynolds number, values equal to Re = 98000, Re = 111000, Re = 124000 and Re = 137000 were considered in this study. To do this, we have used an open wind tunnel to evaluate the global characteristics of the wind turbine. The overall performance evaluation of the rotor was focused on the power, the dynamic and the static torque coefficient evolution.

Keywords: Power, Reynolds number, Savonius rotor, Torque, Wind tunnel.

INTRODUCTION

Renewable energy is obtained from renewable resources like sunlight, wind, rain, tides and waves. Wind power is growing at a fast pace in the world, and wind turbines have been designed to produce electricity using the kinetic energy of the wind. Savonius wind turbines are vertical-axis turbines rotating because of the drag force generated by the blades. Their best features are simplicity, efficiency and very low noise generation. Furthermore, they work at relatively low flow velocities.

Some works have been done to improve the design of the Savonius rotor. For example, Aldos [1] investigated the rise in Savonius rotor power by enabling the blades to move back while on the upwind side. When Cp increases, the power grows by 11.25 percent. Grinspan *et al.* [2], designed a new blade shape with a twist. A maximum power coefficient of 0.5 was obtained. Shikha *et al.* [3], placed a convergent nozzle at the front of the rotor. At lower wind velocities, the nozzle raises the power and stops the negative torque. Menet and Bourabaa [4] examined the effect of some parameters on the characteristics of the flow around a Savonius

---

* **Address correspondence Sobhi Frikha:** Laboratory of Electro-Mechanic Systems, National Engineering School of Sfax, University of Sfax, BP. 1173, Road Soukra km 3.5, 3038, Sfax, Tunisia; E-mail: frikha_sobhi@yahoo.fr

rotor, such as the overlap ratio, the shaft and the Reynolds number. Numerical results were compared to the experimental data given by Blackwell *et al.* [5].

Kamoji *et al.* [6], studied the influence of the Reynolds number on a modified Savonius rotor. The coefficient of power rises by 19% when the Reynolds number increases. Khan *et al.* [7], tested different blades for different values of the overlap. The highest Cp of 0.375 was obtained for a blade profile of the S-section Savonius rotor at an optimum overlap ratio of 30%. Driss and Abid [8] performed a numerical study of the turbulent flow around a Savonius rotor. They compared their numerical results with experimental data. Driss *et al.*, [9] compared various rotor models having different bucket angles. They reported that the depression zones increased with the rise in the angle of the bucket arc. On the convex surface of the bucket, the acceleration zone is formed and gets higher on the bucket arc.

Rogowski and Maroński [10] numerically investigated the aerodynamic performance of the Savonius rotor. The research has shown that the CFD techniques confirm the experimental data and can be used to optimize the shape of the Savonius rotor buckets. Sharma and Gupta [11] have investigated the efficiency of a three-bucket Savonius rotor using Fluent. They analyzed the flow behavior around the rotor in terms of pressure, velocity, and vorticity contours, for various overlap ratios. Choudhury and Saraf [12] studied the flow behavior of two-bladed Savonius rotor by ANSYS Fluent. They reported that the highest values of the drag and the torque coefficients are obtained at $0°$ and $30°$ rotor blade angles, respectively, and that the highest values of the vorticity and the turbulent kinetic energy are obtained at $30°$ rotor angle. In this paper, we have conducted an experimental study to analyze the influence of the Reynolds number on the aerodynamic behavior of the flow.

MATERIAL AND METHOD

The wind rotor is made up of two half-cylindrical buckets in Plexiglas with a diameter of d=100 mm and a height of H=300 mm (Fig. **1**).

EXPERIMENTAL METHODS

The design of the wind tunnel is displayed in (Fig. **2**). It consists of the ventilation chamber (1), the diffuser (2), the test section (3), the collector (4), the plenum (5), the support (6), and the wind turbine (7). The length of the tunnel is equal to 3857 mm. The test section is of 400 mm x 800 mm x 400 mm.

An AM-4204 anemometer has been used to determine the wind speed at various locations (Fig. **3**). In the test section, the maximum air velocity value is 12.7 m.s^{-1}. We have used a digital tachometer CA-27 to measure the rotational speed of the turbine rotor (Fig. **4**). A torque meter TQ-8800 model has been used to measure the static torque on the rotor shaft (Fig. **5**). A DC generator was used to calculate the dynamic torque exerted on the rotor shaft, converting the torque into an electric current. The generator is coupled to the dynamometer RZR-2102 model. In these conditions, the Savonius wind rotor is positioned in the middle of the wind tunnel test section (Fig. **6**).

Fig. (1). Savonius wind rotor.

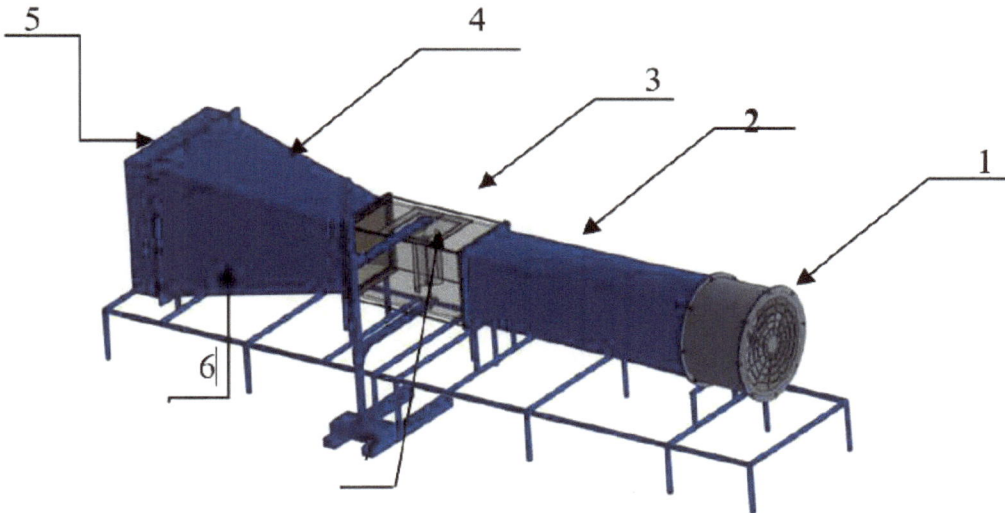

Fig. (2). Wind tunnel components.

Fig. (3). Anemometer.

Fig. (4). Tachometer.

Fig. (5). Torque meter.

Fig. (6). Test vein of the wind tunnel.

EXPERIMENTAL RESULTS

In this work, we are interested in studying the influence of the Reynolds number on the aerodynamic features of the Savonius rotor. By varying the wind speed in the test stream, we studied the flow around the wind rotor for various Reynolds numbers. This can be obtained by a continuous variation of the frequency of the ventilator between 0 Hz and 60 Hz. In this manipulation, four configurations with different values of the frequency are considered equal to f = 30 Hz, f = 35, f = 40 Hz and f = 50 Hz. The corresponding Reynolds number values are Re = 98000, Re = 111000, Re = 124000 and Re = 137000 respectively. In this study, the main recovery is equal to (e-e') / d = 0. For each Reynolds number, the power coefficient C_p and the moment coefficient C_{Md} were determined.

Power Coefficient

Fig. (**7**) displays the variation of the power coefficient C_p as a function of the specific speed λ for the different values of the Reynolds number. In Fig. (**8**) are superposed the data for all the configurations. According to these findings, the evolution of the Cp reveals a parabolic form. This means that a maximum value is to be considered for the rotor characteristics. By varying the Reynolds number from Re = 98000 to Re = 137000, it has been noted that the highest value of the power coefficient increases from a value equal to 0.17 to a value equal to 0.26. Furthermore, the highest value of the power coefficient depends on the Reynolds numbers. For example, for Re = 98000 the highest value corresponds to a specific speed equal to $\lambda = 0.33$. However for a Reynolds number equal to Re = 137000, the highest value is obtained for $\lambda = 0.35$.

Dynamic Torque Coefficient

Fig. (**9**) displays the variation of the torque coefficient C_{Md} as a function of the specific speed λ for the different values of the Reynolds number. Fig. (**10**) shows the data for all the configurations. It has been noted that the highest value of the dynamic torque coefficient depends on the Reynolds number value. Indeed, the torque coefficient values decrease with the decrease of the Reynolds number. This fact is in good agreement with the increase of the power coefficient and the Reynolds number for the same rotor speed. For a Reynolds number equal to Re = 98000, the highest value of the torque coefficient is equal to $C_{Md} = 0.58$. Furthermore, the specific speed λ corresponding to the highest value of the torque coefficient C_{Md} decreases with the Reynolds number. For example, for Re = 98000, the specific velocity is equal to $\lambda = 0.275$.

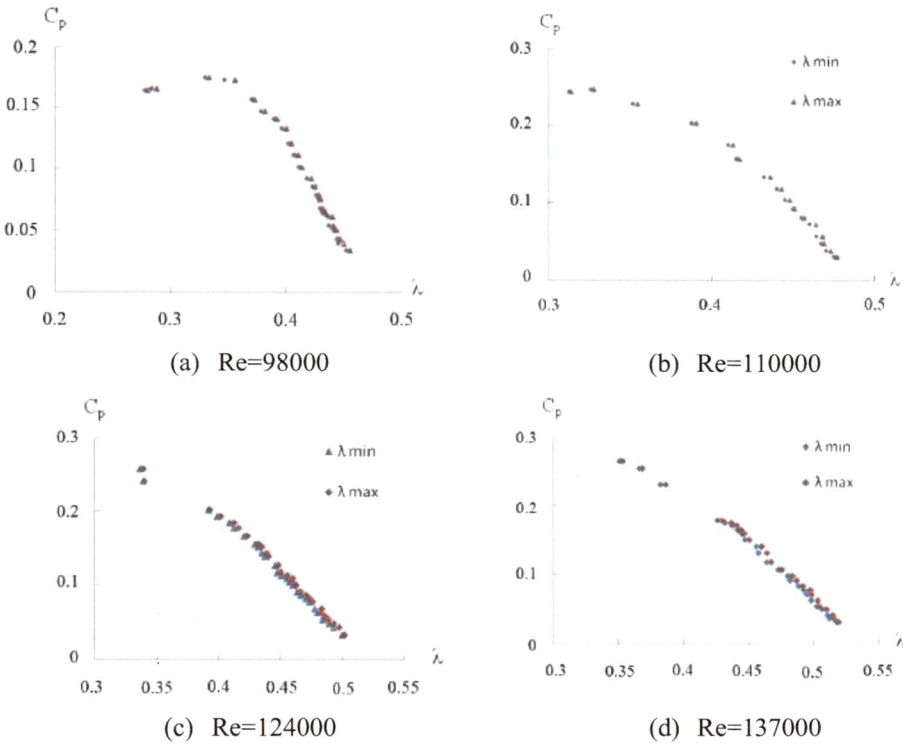

(a) Re=98000

(b) Re=110000

(c) Re=124000

(d) Re=137000

Fig. (7). Variation of the power coefficients.

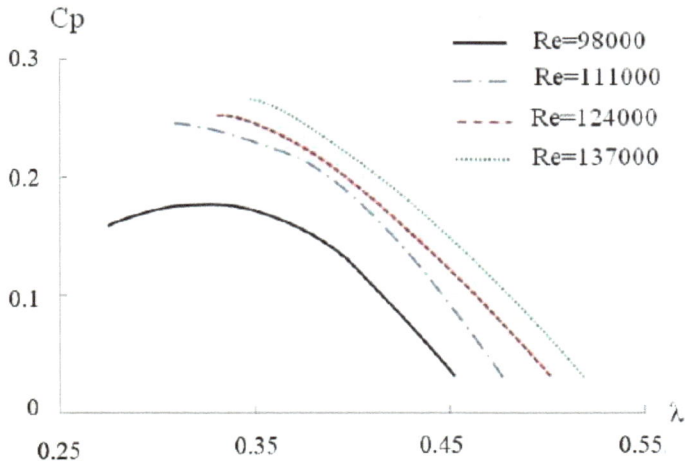

Fig. (8). Comparison of the power coefficient for different Reynolds numbers.

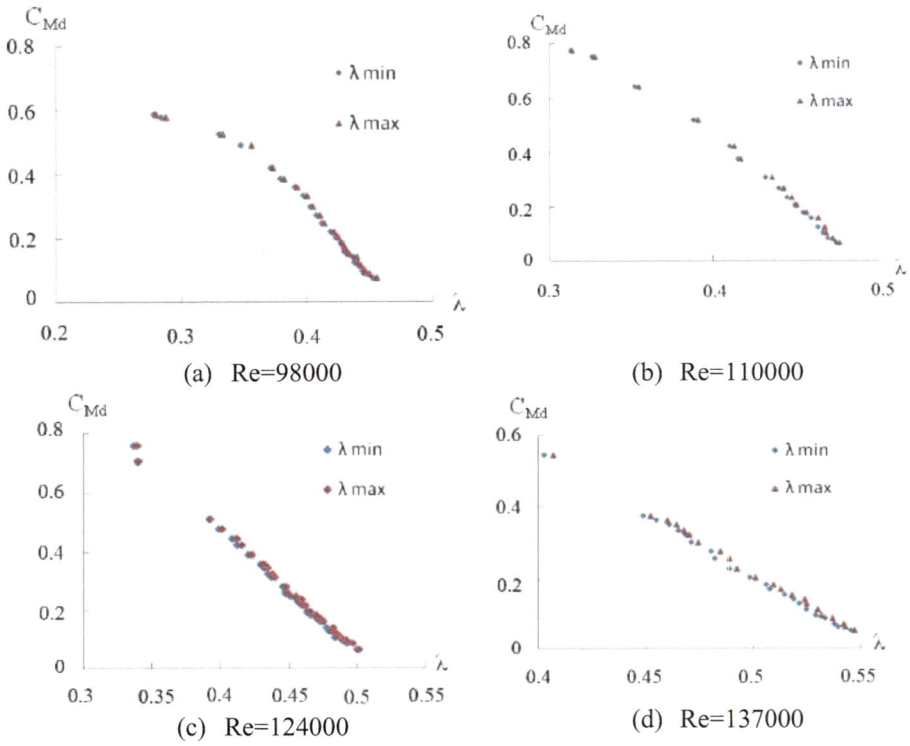

Fig. (9). Variation of the dynamic torque coefficient.

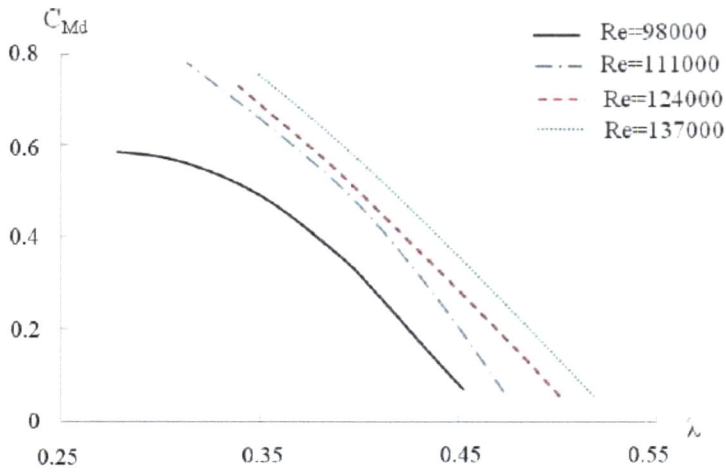

Fig. (10). Comparison of the dynamic torque coefficient for different Reynolds numbers.

Static Torque Coefficient

Fig. (**11**) displays the variation of the static torque coefficient C_{Ms} as a function of the incidence angle Θ of the Savonius rotor for the different values of the Reynolds number. Fig. (**12**) shows the data for all the configurations. According to these findings, the different curves have the same shape. The highest value of the static torque coefficient is given at Re = 137000. It is equal to C_{Ms} = 0.28. In this case, the corresponding incidence angle is equal to Θ = 120 °. The curve presents two other extremums defined for Θ = 30 ° and Θ = 60 °. The values of the corresponding static torque coefficients are equal to C_{Ms} = 0.12 and C_{Ms} = 0.05. Furthermore, the lowest value of the static torque coefficient C_{Ms} is achieved at Re = 98000. In this case, the values of C_{Ms} and the angle of incidence are C_{Ms} = 0.21 and Θ = 30 °, respectively. The curve admits a second extremum equal to C_{Ms} = 0.1 for Θ = 150 °. The two peaks in the C_{Ms} correspond to the number of blades (two) on the Savonius wind rotor.

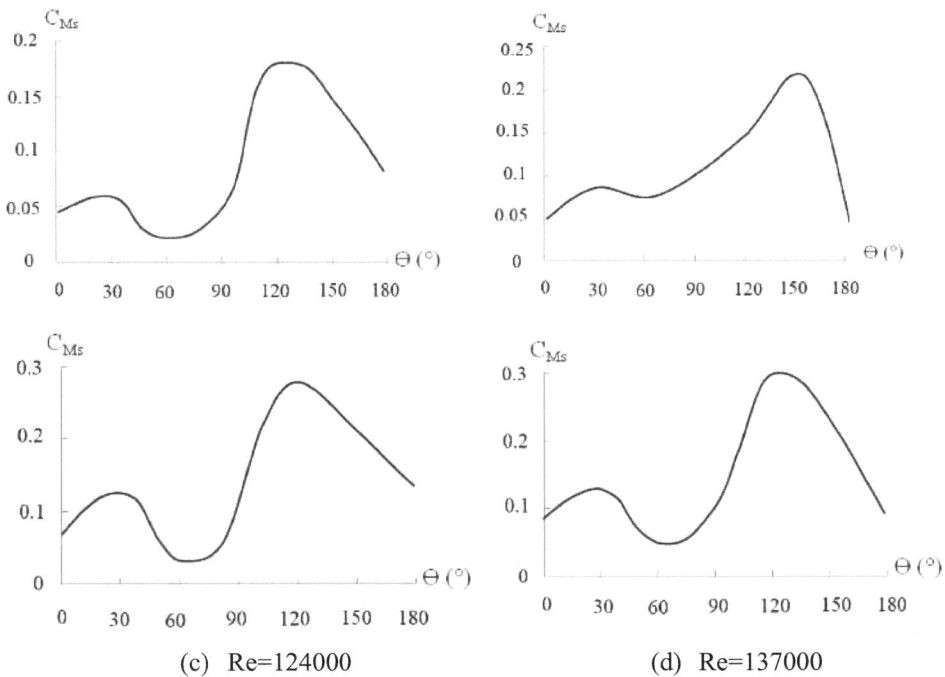

(c) Re=124000   (d) Re=137000

Fig. (11). Variation of the static torque coefficient.

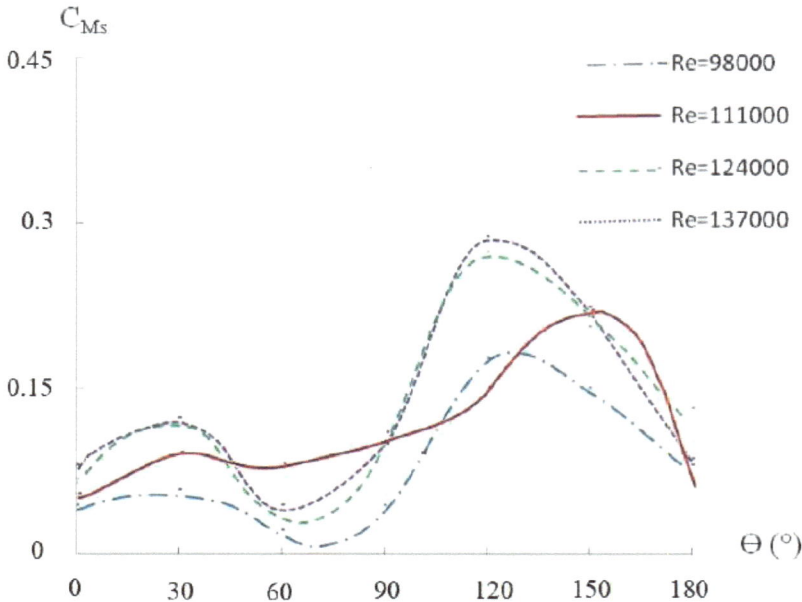

Fig. (12). Comparison of the torque coefficient for different Reynolds numbers.

CONCLUSION

In this paper, we have investigated the global characteristics of a Savonius rotor with various Reynolds numbers. The experiments were performed in an open-air wind tunnel. In particular, the performance of the rotor has been measured based on the power coefficient, the dynamic torque coefficient and the static torque coefficient. It has been shown that the Reynolds number affects the global characteristics of the Savonius wind turbine. Indeed, by varying the Reynolds number from Re = 98000 to Re = 137000, the highest value of the power coefficient increases from 0.17 to 0.26. Furthermore, the highest value of the power coefficient depends on the Reynolds numbers. In addition, the dynamic torque coefficient values decrease with the decrease of the Reynolds number. Furthermore, the specific speed λ corresponding to the highest value of the torque coefficient C_{Md} decreases with the Reynolds number. In addition, the different curves of the static torque coefficient have the same shape. The maximum value of the static torque coefficient is achieved for Re = 137000. Furthermore, the minimum value of the static torque coefficient C_{Ms} is given for Re = 98000.

CONSENT FOR PUBLICATION

Not Applicable.

CONFLICT OF INTEREST

The author declares no conflict of interest, financial or otherwise.

ACKNOWLEDGEMENTS

Declared none.

NOMENCLATURE

A   Rotor area

C$_{MD}$

   Dynamic torque coefficient dimensionless, $C_{Md} = \dfrac{4\,M_d}{\rho V^2 D^2 H}$

C$_{MS}$

   Static torque coefficient dimensionless, $C_{MS} = \dfrac{4\,M_S}{\rho A D V^2}$

C$_p$

   Power coefficient dimensionless, $C_p = \dfrac{2P}{H D \rho V^3}$

d   Rotor diameter, m

H   Bucket height, m

M$_d$ Dynamic torque, N.m

M$_S$ Static torque, N.m

P   Power, W

r   Rotor radius, m

R$_e$

   Reynolds number, dimensionless, $Re = \dfrac{VD}{\nu}$

ν   Kinematic viscosity.

λ

   Tip speed ratio, dimensionless, $\lambda = \dfrac{\Omega r}{V}$

Ω   Angular velocity, rd.s^{-1}

Θ   incidence angle, rd

REFERENCES

[1]   T.K. Aldos, "Savonius rotor using swinging blades as an augmentation system", *Wind Eng.,* vol. 8, pp. 214-220, 1984.

[2]   A.S. Grinspan, S. Kumar, U.K. Saha, P. Mahanta, D.V. Ratnarao, and G. Veda Bhanu, "Design, development & testing of Savonius wind turbine rotor with twisted blades",

[3]   "Shikha, Bhatti TS, Kothari DP. Wind energy conversion systems as a distributed source of

generation", *J. Energy Eng.,* vol. 129, no. 3, pp. 69-80, 2003.
[http://dx.doi.org/10.1061/(ASCE)0733-9402(2003)129:3(69)]

[4]   J.L. Menet, and N. Bourabaa, "Increase in the Savonius rotors efficiency via a parametric investigation", *European wind energy conference,* 2004

[5]   B.F. Blackwell, R.E. Sheldahl, and L.V. Feltz, "Wind Tunnel performance data for two and three-bucket Savonius rotor", *J. Energy,* vol. 2–3, pp. 160-164, 1978.

[6]   M.A. Kamoji, S.B. Kedare, and S.V. Prabhu, "Experimental investigations on single stage modified Savonius rotor", *Appl. Energy,* vol. 86, pp. 1064-1073, 2009.
[http://dx.doi.org/10.1016/j.apenergy.2008.09.019]

[7]   N. Khan, I.M. Tariq, M. Hinchey, and V. Masek, "Performance of Savonius rotor as water current turbine", *J. Ocean Technol.,* vol. 4, no. 2, pp. 27-29, 2009.

[8]   Z. Driss, and M.S. Abid, "Numerical investigation of the aerodynamic structure flow around Savonius wind rotor", *Sci Acad Trans Renew Energy Syst Eng Technol,* vol. 2, no. 2, pp. 196-204, 2012.

[9]   Z. Driss, O. Mlayeh, S. Driss, D. Driss, M. Maaloul, and M.S. Abid, "Study of the bucket desgin effect on the turbulent flow around unconventional Savonius wind rotors", *Energy,* vol. 89, pp. 708-729, 2015.
[http://dx.doi.org/10.1016/j.energy.2015.06.023]

[10]   K. Rogowski, and R. Maronski, "CFD computation of the Savonius rotor", *J. Theor. Appl. Mech.,* vol. 53, no. 1, pp. 37-45, 2015.
[http://dx.doi.org/10.15632/jtam-pl.53.1.37]

[11]   K.K. Sharma, and R. Gupta, "Flow field around three bladed Savonius rotor", *Int J Appl Eng Res,* vol. 8, no. 15, pp. 1773-1782, 2013.

[12]   B.J. Choudhury, and G. Saraf, "Computational analysis of flow around a two-bladed Savonius rotor", *ISESCO J Sci Technol,* vol. 10, no. 17, pp. 39-48, 2014.

CHAPTER 5

Effect of Operating Parameters on Zn-Mn Alloys Deposited from Additive-free Chloride Bath

Nouha Loukil[1,*] and **Mongi Feki**[1]

[1] *Laboratory of Material Engineering and Environment, ENIS-Tunisia, University of Sfax, Sfax, Tunisia*

Abstract: Zn-Mn electrodeposition from additive-free chloride bath on steel was investigated. Several operating parameters, namely the Mn^2 concentration, the current density and the stirring were explored with regard to the Mn content in the Zn matrix. The Mn content depends on the applied current density and jumps from zero to a maximum of 11.4% under 140 mA/cm². At high current density, Zn-Mn coatings are darker, more dendritic and with bad adhesion to the substrate. The dark appearance of Zn-Mn alloys is linked to oxy/hydroxide inclusions formed into the co-deposits.

Keywords: Electrodeposition, Morphology, Oxy/hydroxides, Zn-Mn alloy.

INTRODUCTION

Zn-Mn alloys are currently in demand in the automotive industry due to their high anti-corrosion properties compared to pure Zn [1]. Zn-Mn alloys can offer sacrificial protection to steel in corrosive media (NaCl, SO_2....). Boshkov *et al.* [2] reported that Mn had been suggested due to its dual protective action. Considering that Mn is more electronegative than Zn, Mn dissolves as Mn^{2+} ions, leading to a rise of pH in the interface deposit/aggressive media. Zn then dissolves when Mn is completely dissolved. Due to the alkaline pH value, Zn^{2+} ions directly react with the medium forming zinc hydroxide chlorides $Zn_5(OH)_8Cl_2H_2O$ (ZHC) or zinc hydroxide sulfates $Zn_4(OH)_6SO_44H_2O$ (ZHS) in chloride (5% NaCl) or in sulfate medium (1N Na_2SO_4), respectively. The formed passivation layer is compact with low solubility of $10^{-14.2}$ mol dm^{-3} [3, 4].

Several works studied the effect of the Mn amount on the anticorrosive properties of Zn-Mn alloys. The optimum anti-corrosion behavior of Zn-Mn alloys is obtained with 20% Mn [5]. Zn-Mn electrodeposition from aqueous solution is a

---

* **Corresponding author Nouha Loukil:** Laboratory of Material Engineering and Environment, ENIS-Tunisia, University of Sfax, Sfax, Tunisia; E-mail: nloukil87@gmail.com

Zied Driss (Ed.)

challenging issue due to the great gap between the deposition potentials of Zn ($E°$ (Zn^{2+}/Zn) = -0.76 V/SHE1 and Mn ($E°$ (Mn^{2+}/Mn) = -1.18 V/SHE1). Both potentials of the two alloying elements are more negative than that of hydrogen evolution reaction. This concomitant reaction leads to serious drawbacks, weak current efficiency and poor adhesion of Zn-Mn coatings [6 - 8].

When the standard potentials of two metals are so far, the operating parameters can be adjusted to get closer the deposition potentials by displacing the deposition potential of the nobler metallic element to that of the less noble element. This permits an increase in the amount of the less noble metal in the alloy.

The present work aims to investigate the effects of the operating parameters, namely current density, Mn^{2+} concentration and stirring of the electrolyte on the Mn content and the morphology of Zn-Mn alloys.

MATERIAL AND METHODS

Bulk electrolysis of Zn-Mn alloys was implemented on steel plates with dimensions of 30×40 mm^2 (Fig. **1**) from a simple aqueous chloride bath consisting of $ZnCl_2$, $MnCl_2$, KCl and H_3BO_3. According to the literature [9, 10], a simple chloride electrolyte shows numerous benefits in comparison with other electrolytic bath types (pyrophosphate [11, 12] and sulfate bath [13]). Literature data stated that chloride bath displays high current efficiency (near to 100%) as well as good bath stability.

For these experiments, the steel substrate retained as a cathode and a Zn anode were related to a dc power supply (Fig. **1**). The bulk electrolysis tests were performed for 30 min at room temperature.

The chemical composition of the working solution is described in Table **1**. Prior to all tests, pH was adjusted to 4.5 with hydrochloric acid (HCl) and/or potassium hydroxide solution (KOH).

Table 1. Chemical composition of the electrolytic bath.

Chemical Composition	$ZnCl_2$	$MnCl_2.2H_2O$	KCl	H_3BO_3
Concentration (mol/l)	0.3	0.3 – 1	1.25	0.4

In advance of each test, the substrate surfaces were polished with different grades of abrasive paper to obtain smooth surfaces. Then, the substrate surfaces were degreased in alcohol. Then, they were pickled in a hydrochloric acid medium (10%) for 30 s just before use.

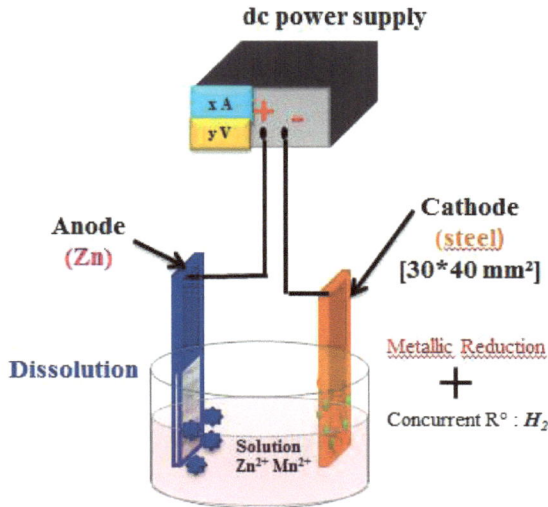

Fig. (1). Bulk electrolysis cell related to a digital dc power supply to deposit Zn-Mn alloy.

Zn-Mn co-deposits were dissolved from the cathode surface in sulfuric acid H_2SO_4 solution (1 M). Hexamethylenetetramine (HMTA) (2 g/l) was used as a dissolution inhibitor for steel substrate. Zn and Mn percentages were measured *via* atomic absorption spectrometry (Analytic Jena ZEENIT 700). This measurement was realized in triplicate. The morphologies of Zn-Mn co-deposits were investigated using a stereomicroscope (Olympus SZ61).

RESULTS AND DISCUSSION

Effect of the Current Density

Referring to voltammogram data described in our previous work dealing with a similar chloride bath (Fig. **2**) [7], Zn^{2+} and Mn^{2+} co-deposition starts at a deposition potential E_d of -1.52 V. This potential is more anodic compared to that recorded from sulfate bath. Indeed, Muller *et al.* [13] established that Mn starts to be co-deposited from sulfate bath only when the deposition potential is more negative than -1.62 V.

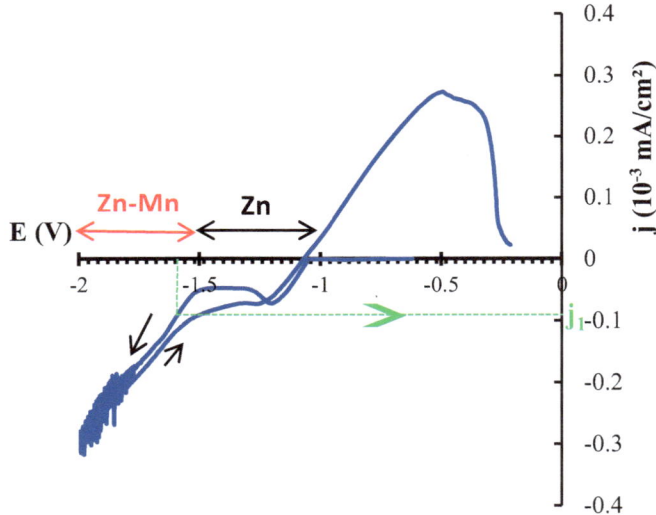

Fig. (2). Determination of the selected current density from the cyclic voltammogram for Zn-Mn deposition [7].

The voltammogram curve (corresponding to the blue line) exhibits a cathodic peak that appears at -1.03 V, followed by a current density plateau. This steady current density corresponds to the limiting current density j_L for Zn deposition (Fig. **2**). The presence of this current density plateau reveals that Zn^{2+} ions reduction occurs according to diffusion control prior to Mn^{2+} reduction [7, 9]. Thus, the applied current densities have to be greater than the limiting current density j_L observed for Zn^{2+} reduction to produce Mn-rich alloys [14]. Fig. (**2**) illustrates how to determine the current density from the voltammogram (blue line). For selected potential superior to -1.52 V, the corresponding current density is determined by projection on the ordinate axis j (mA/cm²). The projection method is designated with a green dotted line (Fig. **2**). Bulk electrolysis experiments were performed at current densities, namely 40, 60, 140 and 200 mA/cm². Fig. (**3**) shows the evolution of the Mn content in Zn-Mn alloys deposited from the acidic bath under various current densities. The Mn amount increases as the current density are higher. Increasing the current density from 40 to 140 mA/cm² induces a significant rise in the Mn content from 0.98 to 11.4%. This is in well agreement with previous data showing that lower cathodic potentials are compulsory to enable the co-deposition of an acceptable Mn content [2, 15]. This additive-free bath seems to be very useful as it allows an easier Zn-Mn co-deposition. Meanwhile, the Mn amount slightly falls down if the current density rises until 200 mA/cm² (Fig. **3**). This decrease is certainly ascribed to an excessive hydrogen ions discharge under high current density. This great extent of hydrogen ions evolving is related to the catalytic activity of Mn-rich alloys [7, 9]. As the Mn amount is greater, the hydrogen evolution reaction is more intensive. Subsequently, a loss of Zn-Mn alloys in the form of very fine particles from the cathode surface is detected, generating a reduction of the weight of Zn-Mn alloys [5]. This assumption explains the decrease of the current efficiency at higher current density (Fig. **3**) .

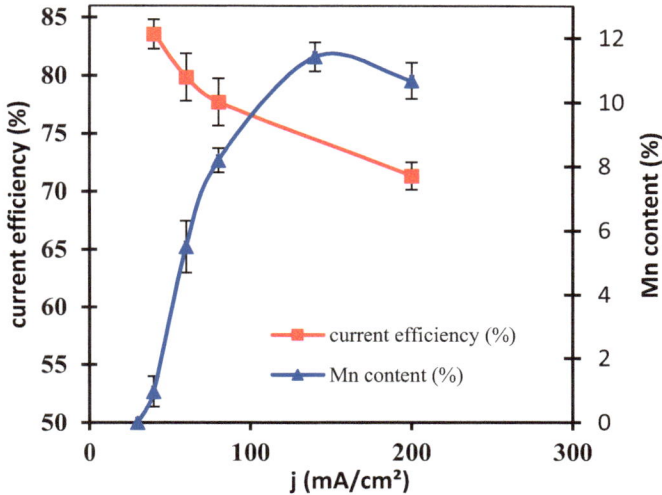

Fig. (3).   Effect of the current density **j** on the Mn content and on the current efficiency of the working electrolytic bath.

Effect of Mn²⁺ Concentration

The effect of Mn^{2+} concentration present in the working electrolytic bath is investigated at a fixed current density (60 mA/cm²). Fig. (**4**) shows the impact of three different Mn^{2+} concentration values on the Mn content. Increasing Mn^{2+} molarity from 0.3 to 1 M induces an improvement of the Mn content that passes from 2.7% to 5.5% (Fig. **4**). This corroborates the basic principles of the normal co-deposition of Zn-Mn [14]. When two alloying elements have different standard potentials, increasing the concentration of the less noble metal element enhances its deposition [14]. The chemical results of the Mn content attest that Zn-Mn alloys can be successfully produced from a simple chloride bath (without any complexing agent). According to literature data, Boshcov [3] stated that it is impossible to co-deposit Zn-Mn alloys from a simple sulfate bath containing a similar Mn^{2+} concentration (0.36 mol/l). It seems that Cl^- ions and H_3BO_3 reduce the potential gap between Zn^{2+} and Mn^{2+} reduction and hinder hydrogen reduction on the cathode [15]. Bucko *et al*. [15] suggested that metal-chloride complex compounds act as an ion bridge between the cathodic surface and the metal ion.

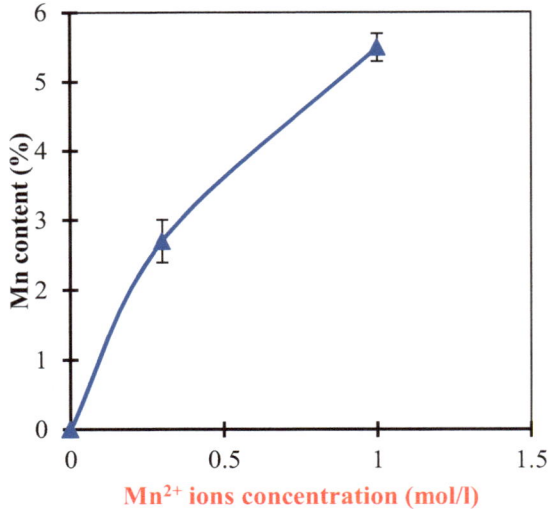

Fig. (4). Effect of Mn^{2+} concentration on the Mn content in Zn-Mn alloys.

Although the Mn^{2+} concentration value of 1 M is much greater, the rise in the Mn content is considered small, not as expected (Fig. **4**). However, the electrodeposition from the electrolyte with the highest Mn^{2+} concentration (1 M) leads to an enhancement in the current efficiency from 65 to 80%. Despite of this advantage, a further rise in Mn^{2+} concentration seems to be not advisable. According to literature data [5, 8], Selvarani *et al.* [5] reported that Mn^{2+} concentration of 2 M lowers the current efficiency of Zn-Mn electrodeposition due to the intensive hydrogen evolution reaction that reduces the final weight of Zn-Mn alloys. In this context, Bucko *et al.* [8] also demonstrated that high Mn^{2+} concentration catalyzes H_2 evolving, diminishing then the current efficiency. Taking into account all these results, a constant Mn^{2+} concentration equal to 1 M was retained for the following experiments.

Effect of Stirring on Mn Content

Fig. (**5**) displays the effect of stirring of the electrolytic solution on the Mn content in Zn-Mn alloys. Stirring the solution is one of the main operating factors that determine the final composition of alloys. For a current density equal to 60 mA/cm^2, the Mn content falls down from 5.5% to 0.3%. Similar results are reported in previous works [9]. Stirring the solution is not beneficial for Mn co-deposition into Zn matrix. It rather enhances Zn electrodeposition due to speeding up of the diffusion-controlled Zn^{2+} reduction [9] or to the removal of hydrogen bubbles from the cathode.

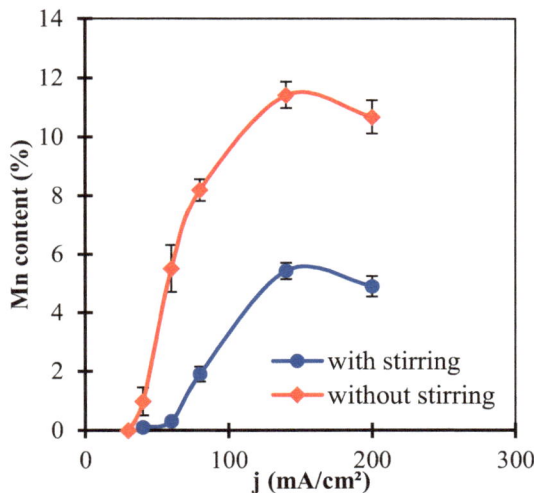

Fig. (5). Effect of stirring of the solution on the Mn percentage at various current densities.

Morphology of Zn-Mn Alloys

Fig. (**6**) exhibits the optical macrographs of Zn-Mn deposits electroplated at various applied current densities. Zn-Mn coatings obtained under the lowest current density (40 mA/cm^2) are gray and slightly dendritic, while those formed under higher current density are darker, spongy and powdery. All the coatings seem to be with very poor adhesion to the steel. Taking into account the chemical composition of co-deposits, this difference in color is absolutely linked to Mn co-deposition in the Zn matrix. These occurrences are in accordance with the literature data [5, 16].

By examining the optical macrographs of Zn-Mn deposits (Fig. **6**), more cathodic deposition potential affects the surface uniformity of Zn-Mn deposits. The surface morphology observed at high current density displays aggregates formed by small nodular grains. As shown in Fig. (**6**), it is important to note that agglomerates are non-uniformly distributed on the entire surface. As a function of the current density, the agglomeration size is proportional to the current density (Fig. **6**). This is linked to both diffusion-controlled Zn^{2+} ions reduction and intensive hydrogen discharge [15]. Indeed, the dendritic morphology is a common characteristic of metallic coating plated at very negative deposition potentials [17]. These observations disclose that Mn co-deposition significantly modifies the morphology of Zn-Mn alloys. These observations are associated with the hydrogen evolution reaction that is very excessive at high current density.

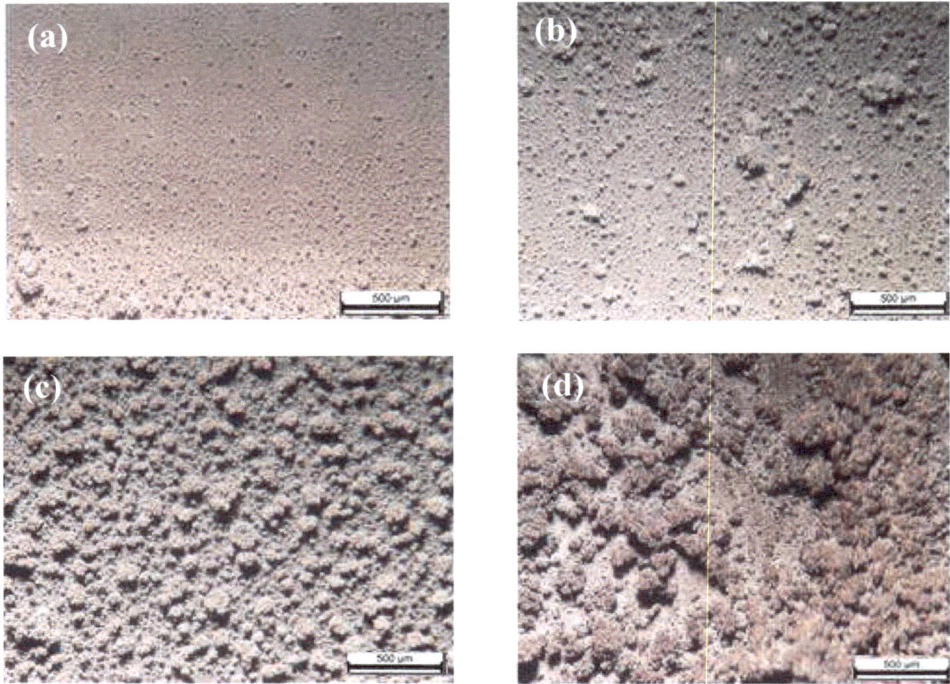

Fig. (6). Optical macrographs of Zn-Mn samples deposited at various current densities **(a)**: 40 mA/cm^2; **(b)**: 60 mA/cm^2, **(c)**: 140 mA/cm^2; **(d)**: 200 mA/cm^2

Some authors proposed that hydrogen presence alters the surface energy and growth mechanisms, increasing then the agglomerate size at high current density [15 - 17]. Due to hydrogen evolution, (OH$^-$) are excessively present in the near-cathodic layer, leading to an increase of pH in the metal/solution interface [15, 19]. This increase in pH leads to the formation of oxy/hydroxides Zn(OH)$_2$ and/or Mn(OH)$_2$. These oxy/hydroxides lead to the dark appearance of Zn-Mn alloys [19].

CONCLUSION

All results show that an additive-free chloride bath permits to produce Zn-Mn alloys when the current densities are greater than the limiting current density for Zn deposition. Applying high current density induces Mn-rich alloy formation. However, this enhancement of Mn content takes place in the detriment of the current efficiency. Indeed, Mn-rich alloys catalyze the hydrogen evolution reaction, resulting not only dendritic morphology, but also metallic hydroxides precipitation in Zn-Mn alloys. The increase of Mn^{2+} concentration is useful for easier co-deposition of Mn, but a strong increase of Mn^{2+} concentration has a reverse effect on the current efficiency. In contrast, stirring the electrolyte has a common effect to decrease the Mn content.

NOTES

[1] SHE: Saturated Hydrogen Electrode

CONSENT FOR PUBLICATION

Not Applicable.

CONFLICT OF INTEREST

The author declares no conflict of interest, financial or otherwise.

ACKNOWLEDGEMENTS

Declared none.

REFERENCES

[1]   G.D. Wilcox, and D.R. Gabe, "Electrodeposited Zinc Alloy Coatings", *Corros. Sci.,* vol. 35, pp. 1251-1258, 1993.
[http://dx.doi.org/10.1016/0010-938X(93)90345-H]

[2]   N. Boshkov, "Galvanic Zn-Mn alloys-electrodeposition, phase composition corrosion behavior and protective ability", *Surf. Coat. Tech.,* vol. 172, pp. 217-226, 2003.
[http://dx.doi.org/10.1016/S0257-8972(03)00463-8]

[3]   N. Boshkov, S. Vitkova, and K. Petrov, "Corrosion products of zinc-manganese coatings: part I-investigations using microprobe analysis and X-ray diffraction", *Met. Finish.,* vol. 99, p. 56, 2001.
[http://dx.doi.org/10.1016/S0026-0576(01)81437-9]

[4]   N. Boshkov, K. Petrov, S. Vitkova, and G. Raichevsky, "Galvanic alloys Zn–Mn composition of the corrosion products and their protective ability in sulfate containing medium", *Surf. Coat. Tech.,* vol. 194, pp. 276-282, 2005.
[http://dx.doi.org/10.1016/j.surfcoat.2004.09.016]

[5]   S. Ganesan, P. Ganesan, and N. Branko, "Popov. Electrodeposition and characterization of Zn☐Mn coatings for corrosion protection", *Surf. Coat. Tech.,* vol. 238, pp. 143-151, 2014.
[http://dx.doi.org/10.1016/j.surfcoat.2013.10.062]

[6]   P. Díaz-Arista, Z.I. Ortiz, H. Ruiz, R. Ortega, Y. Meas, and G. Trejo, "Electrodeposition and characterization of Zn–Mn alloy coatings obtained from a chloride-basedacidic bath containing ammonium thiocyanate as an additive", *Surf. Coat. Tech.,* vol. 203, pp. 1167-1175, 2009.
[http://dx.doi.org/10.1016/j.surfcoat.2008.10.015]

[7]   N. Loukil, and M. Feki, "Synergistic effect of triton X100 and 3-hydroxybenzaldehyde on Zn-Mn electrodeposition from acidic chloride bath", *J. Alloys Compd.,* vol. 719, pp. 420-428, 2017.
[http://dx.doi.org/10.1016/j.jallcom.2017.05.142]

[8]   M. Bucko, J. Rogan, B. Jokic, M. Mitric, U. Lacnjevac, and J.B. Bajat, "Electrodeposition of Zn-Mn alloys at high current densities from chlorideelectrolyte", *J. Solid State Electrochem.,* vol. 17, pp. 1409-1419, 2013.
[http://dx.doi.org/10.1007/s10008-013-2004-8]

[9]　D. Sylla, J. Creus, C. Savall, O. Roggy, M. Gadouleau, and P. Refait, "Electrodeposition of Zn–Mn alloys on steel from acidic Zn–Mn chloride solutions", *Thin Solid Films,* vol. 424, pp. 171-178, 2003.
[http://dx.doi.org/10.1016/S0040-6090(02)01048-9]

[10]　M.V. Tomic, M.M. Bucko, M.G. Pavlovic, and J.B. Bajat, "Corrosion stability of electrochemically deposited Zn-Mn alloy coatings", *Contemp. Mater.,* vol. I–1, pp. 87-93, 2010.
[http://dx.doi.org/10.5767/anurs.cmat.100101.en.087T]

[11]　D. Sylla, C. Rebere, M. Gadouleau, C. Savall, J. Creus, and P.H. Refait, "Electrodeposition of Zn-Mn alloys in acidic and alkaline baths. Influence of additives on the morphological and structural properties", *J. Appl. Electrochem.,* vol. 35, pp. 1133-1139, 2005.
[http://dx.doi.org/10.1007/s10800-005-9001-2]

[12]　D. Sylla, C. Savall, M. Gadouleau, C. Rebere, J. Creus, and P. Refait, "Electrodeposition of Zn–Mn alloys on steel using an alkaline pyrophosphate-based electrolytic bath", *Surf. Coat. Tech.,* vol. 7, pp. 2137-2145, 2005.
[http://dx.doi.org/10.1016/j.surfcoat.2004.11.020]

[13]　C. Muller, M. Sarret, and T. Andreu, "Electrodeposition of Zn-Mn alloys at low current densities", *J. Electrochem. Soc.,* vol. 149, pp. 600-606, 2002.
[http://dx.doi.org/10.1149/1.1512668]

[14]　A. Brenner, *Electrodeposition of Alloys.* vol. Vol. I & II. Acad Press: New York, London, 1963, pp. 315-336.
[http://dx.doi.org/10.1016/B978-1-4831-9807-1.50022-2]

[15]　M. Bučko, J. Rogan, S.I. Stevanović, S. Stanković, and J.B. Bajat, "The influence of anion type in electrolyte on the properties of electrodeposited ZnMn alloy coatings", *Surf. Coat. Tech.,* vol. 228, pp. 221-228, 2013.
[http://dx.doi.org/10.1016/j.surfcoat.2013.04.032]

[16]　J. Gong, and Z. Giovanni, "Electrodeposition of sacrificial tin–manganese alloy coatings", *Mater. Sci. Eng. A,* vol. 344, pp. 268-278, 2003.
[http://dx.doi.org/10.1016/S0921-5093(02)00412-4]

[17]　A.M. Rashidi, and A. Amadeh, "The effect of current density on the grain size of electrodeposited nanocrystalline nickel coatings", *Surf. Coat. Tech.,* vol. 202, pp. 3772-3776, 2008.
[http://dx.doi.org/10.1016/j.surfcoat.2008.01.018]

[18]　V.D. Jović, B.M. Jović, and M.G. Pavlović, "Electrodeposition of Ni, Co and Ni–Co alloy powders", *Electrochim. Acta,* vol. 51, pp. 5468-5477, 2006.
[http://dx.doi.org/10.1016/j.electacta.2006.02.022]

[19]　F. Xu, Z. Dan, W. Zhao, G. Han, Z. Sun, K. Xiao, and N. Duan, "Electrochemical analysis of manganese electrodeposition and hydrogen evolution from pure aqueous sulfate electrolytes with addition of SeO2", *J. Electroanal. Chem.,* vol. 741, pp. 149-156, 2015.
[http://dx.doi.org/10.1016/j.jelechem.2015.01.027]

CHAPTER 6

Optical Properties and Stability of a Blue-Emitting Phosphor $Sr_2P_2O_7$:Eu^{2+} Under UV and VUV Excitation

Mouna Derbel[1,*] and **Aïcha Mbarek**[1,2]

[1] *Advanced Materials Laboratory, National School of Engineers of Sfax, University of Sfax, BP W 3038, Sfax, Tunisia*

[2] *Blaise Pascal University, Institute of Chemistry of Clermont-Ferrand, UMR 6296 CNRS, BP 10448, 63000 Clermont-Ferrand, France*

Abstract: In this book chapter, divalent europium-activated alpha-distrontium diphosphate (α-$S_2P_2O_7$) phosphors powders were successfully prepared by a conventional solid state reaction method under reduced atmosphere. Synthesized samples were characterized by means of X-ray diffraction (XRD) patterns, nuclear magnetic resonance (NMR) and infrared (IR) spectroscopy which signify the formation of pure single phase of $Sr_2P_2O_7$. The optical properties were studied in both ultraviolet (UV) and vacuum ultraviolet (VUV) regions. The emission spectra were obtained by excitation excited at 131 or 320 nm present a single intense blue-emitting band from 350 nm to 500 nm due to the 5d-4f transition of Eu^{2+}, indicating that α-$Sr_2P_2O_7$:Eu^{2+} phosphor powders are suitable for near-UV light-emitting-diode (LED) chips (360-400 nm). The influence of temperature on the luminescence intensity of α-$Sr_2P_2O_7$:1%Eu^{2+} was investigated. The activation energy (E_a) for thermal quenching was reported. The phosphor shows excellent thermal stability on temperature quenching. The luminescence properties show that this host material has a highly promising blue-emitting phosphor for white-LED applications.

Keywords: Pyrophosphates, Eu^{2+}, Photoluminescence, Phosphors, Light-emitting diodes.

INTRODUCTION

Recently, solid state lighting source used white-emitting diode (LED) have been subject to increasing interest due to high luminous efficiency, long persistence, environmental friendliness, small volume, energy savings and lack of toxic mercury [1, 2]. In order to generate white light, a new approach using combined red, green and blue emitting phosphors has been suggested. With this method, it is

---

* **Corresponding author Mouna Derbel:** Advanced Materials Laboratory, National School of Engineers of Sfax, University of Sfax, BP W 3038, Sfax, Tunisia; E-mail: mounaderbel@gmail.com

Zied Driss (Ed.)

relatively simple to get a higher efficiency white LED and reduce the manufacturing cost [3].

At present, the Eu^{2+} ion is frequently used as an activator in the phosphor with the broad-bands excitation and emission due to parity allowed $4f^7$-$4f^65d$ transitions. The emission band of Eu^{2+}, which is dependent greatly on the host lattice, can be tuned from ultraviolet to red region. Therefore, many Eu^{2+}-activated blue-emitting phosphors for near-UV LED applications have been developed, such as $Na_3RbMg_7(PO_4)_6$:Eu^{2+} [4], $BaCa_2MgSi_2O_8$:Eu^{2+} [5],SrB_2O_4:Eu^{2+} [6].

The development for research of Eu^{2+} ions in different hosts excited by vacuum ultra- violet (VUV, $\lambda<$ 200 nm) radiations has attracted great interest. This is due to study the high-energy states of Eu^{2+} ions and to explore efficient VUV excited luminescent materials, which could be used in mercury-free lamps and plasma display panels (PDPs) [7]. As an important family of luminescent materials, phosphates phosphors have been developed for application in white-light LEDs because of their exceptional properties, for example, the high thermal and chemical stability, moderate phonon energy, the high thermal and chemical stability, the large band gap and the high absorption in VUV region, and the exceptional optical damage threshold [8]. Among these phosphors, Eu^{2+}-alkaline earth metal pyrophosphtes with general chemical formula of $A_2P_2O_7$ and ABP_2O_7 (Aand B are divalent cations) were found to be efficient phosphors in the blue region and thus have been considered as the phosphor component for White light emission diodes *e.g.*, $SrZnP_2O_7$:Eu^{2+} [9], $SrCdP_2O_7$:Eu^{2+} [10], $Ba_2P_2O_7$:Eu^{2+} [11, 12], $Ca_2P_2O_7$:Eu^{2+} [12]. A literature search reveals that Eu^{2+}-doped $Sr_2P_2O_7$ is an efficient blue phosphor utilized in lamps for phototherapy [13, 14]. Furthermore, the influence of auxiliary codopants (*i.e.*, R^{3+} or M^{2+}) on $Sr_2P_2O_7$ was systematically and comparatively studied [13 - 20]. However, to the best of our knowledge, only few studies are concerned by photoluminescence properties of Eu^{2+} -doped pyrophosphates on VUV region.

In this work, the solid-state reaction under a reductive atmosphere is applied to prepare the blue emitting phosphor divalent europium-activated alpha-distrontium diphosphate α-$Sr_2P_2O_7$:Eu^{2+}. The crystal structure, the VUV-UV photoluminescence spectra, and the thermal stability of the phosphor have been presented and investigated in detail.

EXPERIMENTAL

Sample Synthesis

Powder samples with the general formula α-$Sr_{2-x}P_2O_7$:xEu^{2+}(x = 0, 1 and 5 mol %) were synthesized by a conventional solid-state reaction method under reductive atmosphere. Analytical grade $SrCO_3$ (Aldrich, 99.9%), $(NH_4)H_2PO_4$ (Acrõs Organics, 99+%) and Eu_2O_3 (Aldrich, 99.99%) were used as starting materials. Thereafter, the mixtures were preheated under air atmosphere at 400°C for 4h for ejection of volatile gas and then annealed at 900°C for 5h under reducing atmosphere ($5\%H_2+95\%Ar$). Finally, the samples were cooled down to room temperature and pulverized for further characterizations.

Characterization Technique

The X-ray powder diffraction (XRD) patterns were measured with a PhilipsX-Pert Pro diffractometer operating with Cu-Kα radiation ($\lambda=1.5406Å$). ^{31}P magic angle spinning nuclear magnetic resonance (MAS NMR) was carried out at 121.5 MHz with a Bruker Advance 300 spectrometer operating in a static field of 7.1 T. The infrared absorption spectra were recorded with a Nicolet 5SX-FTIR spectrometer equipped with diamond micro-ATR accessory and working with OMNIC software. The UV–visible absorption spectra of the samples were recorded in the wavelength range of 200 to 700 nm with a UV–vis spectrophotometer (SP-3000Plus). The photoluminescence spectra in the range UV-visible were measured using a Jobin-Yvon setup consisting of a Xe lamp operating at 400 W. For the temperature measurement, the setup was equipped with a homemade heating cell connected to a temperature controller. The VUV excitation spectra and corresponding emission spectra were recorded at room temperature using specific system built by Jobin-Yvon, allowing excitation in the range 112–300 nm.

RESULTS AND DISCUSSION

Phase Analysis

Fig. (**1**) shows the typical X-ray diffraction (XRD) patterns of α-$Sr_{2-x}Eu_xP_2O_7$ (x = 0; 0.01; 0.05) phosphors. All diffraction peaks can be identified as diffractions from the α-$Sr_2P_2O_7$ lattice (JCPDS file 24-1011) [21], indicating that the doping concentration range of Eu^{2+} ions does not cause any significant change in the host lattice or do not introduce any impurity phases.

The α-$Sr_2P_2O_7$ pyrophosphate has an orthorhombic crystal system with Pnma space group and the structure contains ecliptic diphosphate groups (m symmetry). [21]. Fig. (**2**) illustrates the crystal structure of α-$Sr_2P_2O_7$. In this structure, two types of PO_4 tetrahedrons are linked to each other by corner sharing and the P_2O_7 groups are isolated by Sr^{2+}. There are two inequivalent Sr^{2+} sites bonded in a 9-coordinate geometry to nine O^{2-} atoms. The two crystallographically different SrO_9 polyhedra are very similar. There are two inequivalent P^{5+} sites bonded to four O^{2-} atoms to form corner-sharing PO_4 tetrahedra. Therefore the doped Eu^{2+}

are expected to occupy the sites of Sr^{2+} ions.

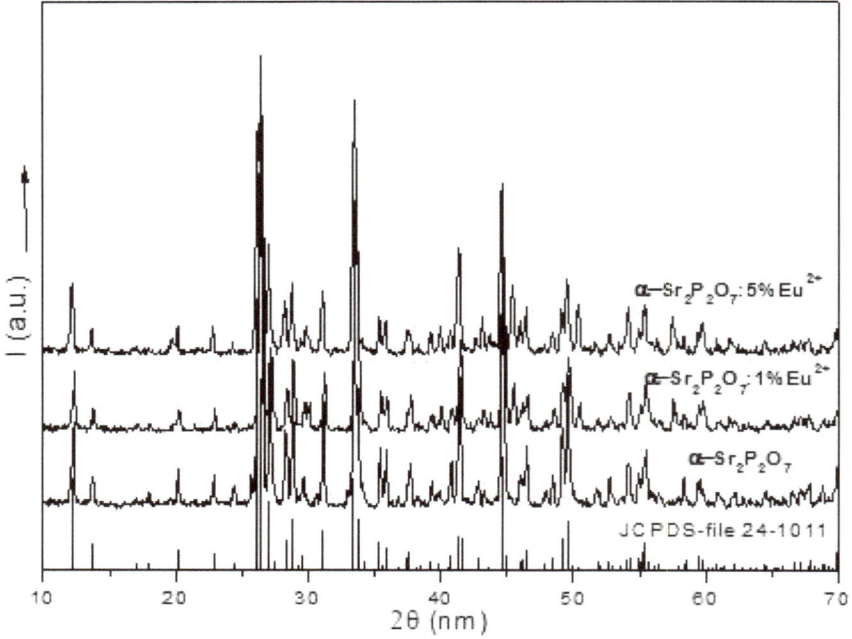

Fig. (1). XRD powder patterns of pure and Eu^{2+}-doped α-$Sr_2P_2O_7$ pyrophosphates.

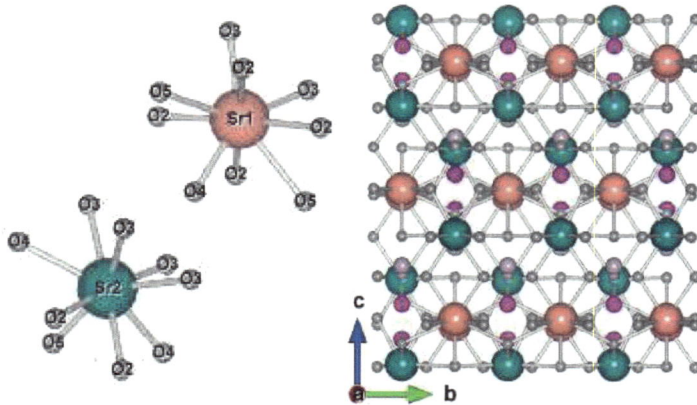

Fig. (2). Schematic views of the α-$Sr_2P_2O_7$ structure along a-direction. Coordination environment around Sr(1) and Sr(2).

Solid-state NMR Spectroscopy

The structure of condensed phosphate can be represented by Q^n nomenclature, where n is the number of bridging oxygen atoms existing on each PO_4 tetrahedron. Especially, pyrophosphates contain Q^1 units, such are phosphorus atoms single-fold coordinated to neighbouring phosphorus atoms *via* oxygen. The Q^1 type units cause the chemical shift in the -10/-30 ppm range [22, 23]. The ^{31}P MAS-NMR solid spectrum, recorded at room temperature, of α-$Sr_2P_2O_7$ is presented in Fig. (**3**). The exploration of the NMR spectrum of this compound exhibit broad band, which contains two lines that appear at -10.69 and -8.09 ppm expected for Q^1 type phosphate environments. This result approves the existence of two crystallographical distinct phosphor sites with two different PO_4 tetrahedra. This is in agreement with the crystal structure, which has two inequivalent phosphorus sites [21].

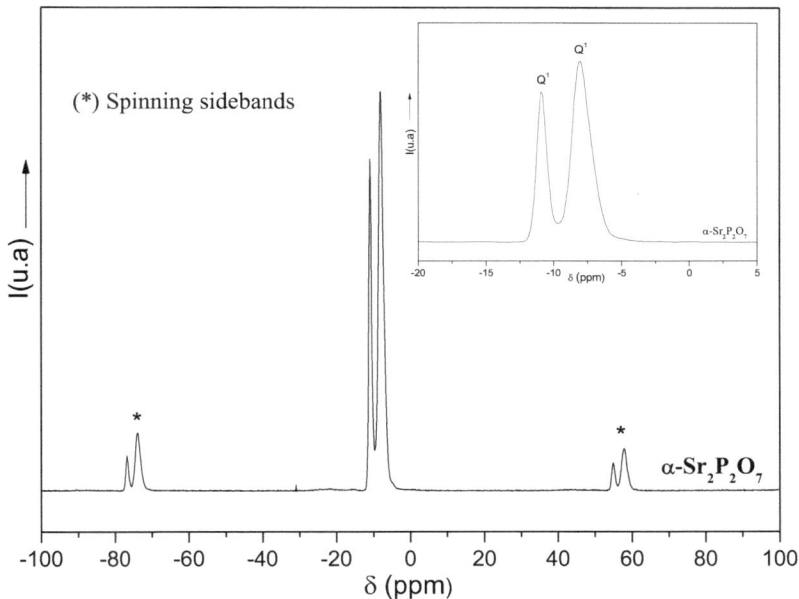

Fig. (3). ^{31}P MAS NMR spectra of $Sr_2P_2O_7$ pyrophosphate.

IR Spectroscopy

The infrared spectra of pyrophosphates offer details concerning the external and modes internal of $P_2O_7^{4-}$ group and metal-oxygen vibrations. Infrared spectra of α-$Sr_2P_2O_7$:Eu^{2+} pyrophosphates samples are reported in Fig. (**4**). These spectra match very well and confirm that no structural modifications are observed when the Eu ions are introduced in the pyrophosphate α-$Sr_2P_2O_7$ matrix. These results are in good agreement with the XRD measurements. The infrared band assignment was carried out by comparison with other pyrophosphates materials [24, 25]. The bands due to the symmetric and antisymmetric stretching frequencies of PO_3 in $P_2O_7^{4-}$ are generally observed in the regions 1200 to 1000 cm^{-1} [26]. The bands observed in the regions 985–800 and 780–690 cm^{-1} are attributed respectively to the antisymmetric and symmetric POP stretching modes [27]. The bands due to $\delta(OPO)$, $\delta(PO_3)$ and $\delta(POP)$ deformations are also identified. It is very difficult to assign precisely these modes because they can also overlap Sr–O or Eu-O stretching vibration frequencies and some external modes in the lower frequencies.

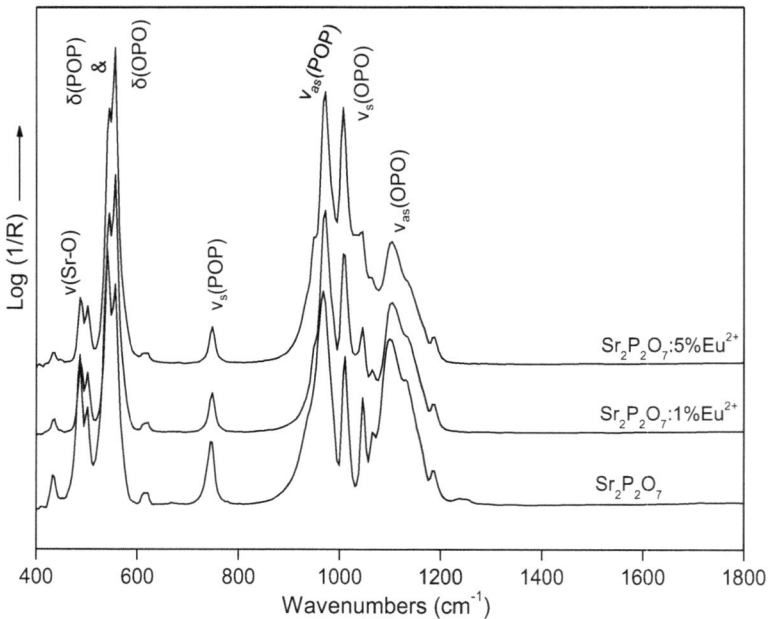

Fig. (4). IR/ATR spectra of α-$Sr_2P_2O_7$:Eu pyrophosphates.

The Absorption of α-$Sr_2P_2O_7$

In order to determine the value of the band- gap, absorption measurements were carried out on α-$Sr_2P_2O_7$ pyrophosphate powder. The absorption spectrum, recorded at room temperature in a wavelength range between 200 and 700 nm, is shown in Fig. (**5a**). Strontium pyrophosphate powder exhibit strong absorption in the ultraviolet range. This corresponds to electronic transitions from the conduction band to the valence band of the host material. The optical absorption of a semiconductor could be calculated by the following Tauc's relation [28]:

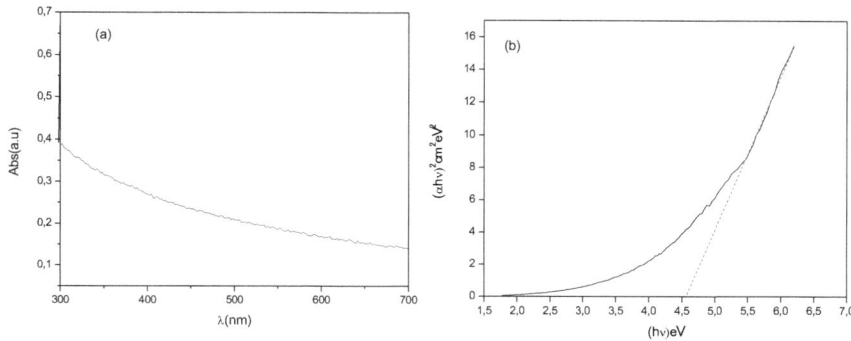

Fig. (5). (a) UV–vis. Absorption spectrum of the α-Sr$_2$P$_2$O$_7$ sample **(b)** Tauc's plot of (αhυ)^2*versus* (hυ) .

$$(\alpha h\nu)^2 = B^2(h\nu - Eg) \qquad (1)$$

where α represents the absorption coefficient, hv is the photon energy, B represents the absorption constant, Eg is the optical band gap.

Fig. (**5b**) depicts the plot of (αhv)^2 along Y-axis and hv along X-axis. The direct band gap of the Sr$_2$P$_2$O$_7$ sample was determined by extrapolating the fitted linear region (αhv)^2 = 0 on X-axis, and found to be 4.55 eV, displaying that this host is a very suitable matrix for rare earth luminescence.

PHOTOLUMINESCENCE PROPERTIES

UV- Vis SPECTROSCOPY

The normalized excitation spectra of Sr$_2$P$_2$O$_7$:Eu^{2+}(x=0.01, 0.05) monitored at 413 nm are shown in Fig. (**6**). A broad excitation band in range 240-440 nm is observed. This symmetric band can be assigned to parity allowed 4f^{6}5d^1→4f^7(^{8}S$_{7/2}$) transition of Eu^{2+} ion situated at Sr^{2+} sites state. It indicates that the phosphor is suitable as a blue phosphor excited by NUV LED chip and mixed with other color emission phosphors to obtain white light.

The emission spectra of α-Sr$_2$P$_2$O$_7$: (1%, 5%) Eu^{2+}are shown in Fig. (**7**). The emission spectra show bright blue luminescence with a peak at 417 nm, attributed to the 4f^{6}5d^1→ 4f^{7}transition of Eu^{2+}. No characteristic peaks of the Eu^{3+} are observed. Apparently, the Eu^{3+} ions were reduced to Eu^{2+} totally by reducing atmosphere (H$_2$/Ar) in our conventional solid-state reaction

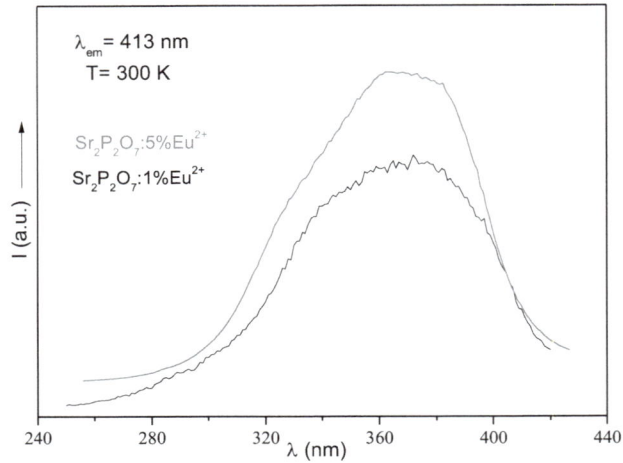

Fig. (6). Excitation spectra of α-$Sr_2P_2O_7$: xEu^{2+} (x = 0.01, 0.05) recorded at 300 K, by monitoring the emission wavelength at 413 nm.

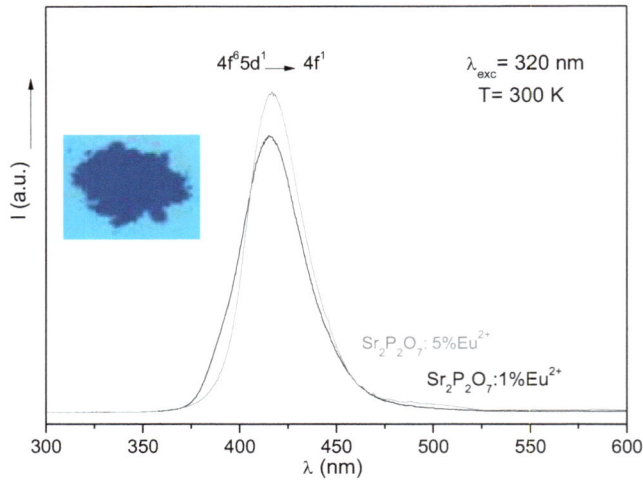

Fig. (7). Emission spectra of α-$Sr_2P_2O_7$:Eu^{2+} (1%, 5%) recorded at 300 K, upon UV excitation at 320 nm.

In order to clarify the assignment of the emission center at 417 nm of Eu^{2+} ions in a host matrix, the empirical equation proposed by Van Uitert [29] has been used to qualitatively analyze the current result. According to this model, the position of the d-band edge in energy of rare earth ions(E) (cm^{-1}), including Eu^{2+}, can be calculated with the help of the following relation:

$$E = Q\left[1 - \left(\frac{V}{4}\right)^{1/V} 10^{-(n.Ea.r)/80}\right]$$

(2)

Where Q represents the position in energy for the lower d-band edge for the free ion (34 000 cm^{-1} for Eu^{2+}), V is the valence state of the activator ion (for Eu^{2+}, V= 2), n is the coordination number of the "active" ion (Eu^{2+}), r is the ionic (in Å) radius of the host cation (Sr^{2+}) replaced by the Eu^{2+} ion and Ea is the electron affinity of the atoms that form the anions (in eV). Here, Ea is determined as 2.19, like reported in other phosphate structures [30, 31]. The calculated energy position of the Eu^{2+} d-band edge in the nine-coordinated Sr^{2+} site (r_i=1.26 Å, CN=9) of Sr$_2$P$_2$O$_7$ is 22237cm^{-1} which is equivalent to the wavelength of 450 nm. As could be seen, experimental values agree well with the calculated values. The result verifies that the Eu^{2+} center that shows blue luminescence does indeed occupy the nine-coordinate Sr^{2+} site.

VUV Spectroscopy

The VUV excitation spectra of α-Sr$_2$P$_2$O$_7$: Eu^{2+} (1%, 5%), obtained by monitoring the blue emission at 417 nm are presented in Fig. (**8**). These spectra are constituted of several bands peaking at about 131 nm, 163 nm, 183 nm and broad absorption in the range 250–400 nm. The two bands appearing at around ~131 and ~163 nm corresponds to the band-to-band excitation of the host crystal [32], *viz* the electrons are promoted from the valence band to the conduction band for the α-Sr$_2$P$_2$O$_7$ host lattice. It is well known that in the structure of α-Sr$_2$P$_2$O$_7$ [21], there are two different (P$_2$O$_7$)$^{4-}$ anionic groups which can explain the two absorption bands displays in Fig. (**8**). The band peaking at ~183 nm is related to the charge transfer band (CTB) of Sr^{2+}-O^{2-} [33]. The excitation band with the maximum at 228 nm could be assignable to charge transfer (CT) absorption of Eu^{2+}-O^{2-} [34]. The broadband within 250-400 nm corresponds to the characteristic electron transition of Eu^{2+} ions between the ground state and the crystal field splitting 5d levels [33].

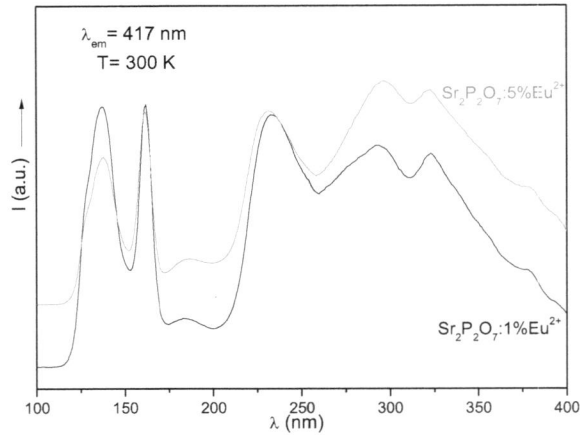

Fig. (8). VUV excitation spectra of α-Sr$_2$P$_2$O$_7$: xEu^{2+} (x = 0.01, 0.05) under λ_{em} = 417 nm.

The VUV emission spectra of the Sr$_2$P$_2$O$_7$:Eu^{2+} (1%, 5%) phosphors under excitation at 163 nm and 131 nm (the host-related absorption) are shown in Fig. (**9**). These spectra are constituted of a broad blue band with a maximum at 417 nm corresponding to the electronic transition of Eu^{2+} from the 4f^{6}5d excited state to the 4f^7 ground state.

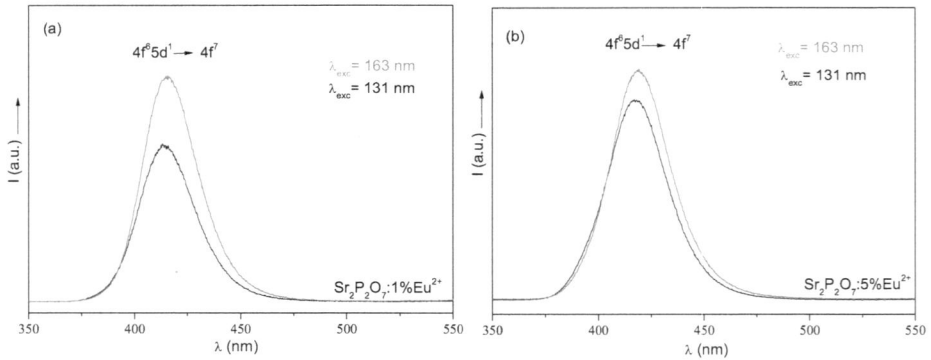

Fig. (9). VUV emission spectra of α-Sr$_2$P$_2$O$_7$: xEu^{2+} (x = 1% (a), 5%(b)) under 163 nm and 131 nm.

Thermal Stability

Thermal stability is one of the important technological parameters to be considered for phosphor-based white LEDs because it has considerable influence on the light output and color rendering index [35, 36]. Fig. (**10**) presents the variation of emission intensity and FWHM of α-Sr$_2$P$_2$O$_7$: 1%Eu^{2+} phosphor excited by 320 nm in the temperature range 25°C–230°C. The temperature dependence of emission intensities is also displayed in the inset of (Fig. **10**). It can be noted from the figure that the relative emission intensity decreases slowly with the increasing temperature. FWHMS increase from 42 nm to 55nm. The increase in FWHMs and the decrease in emission intensities can be investigated by the thermal quenching between the ground state and the excited state in the configurational coordinate diagram [37, 38]. As the temperature increases, the interaction of electron-phonon is intensive. Along with enhancing phonon interaction, more electrons can be thermally activated to the crossover between the 4f^{6}5d excited state and 4f^7 ground state, as a consequence of which release the energy by generating lattice vibration [39].

At 150 °C, the emission intensity of α-Sr$_{0,99}$Eu$_{0,01}$P$_2$O$_7$ phosphor decrease about 67% of its initial value. This result suggests that the phosphor has excellent thermal stability. To facilitate and explain the temperature-dependent thermal quenching phenomenon, the activation energy for the thermal quenching of Eu^{2+} emission has been estimated using the modified Arrhenius equation [40].

$$I_T = \frac{I_0}{1 + c\, \exp\left(-\left(\frac{Ea}{k_B T}\right)\right)} \tag{3}$$

where I_0 is the initial intensity, I_T the intensity at a given temperature T, c a constant and k_B Boltzmann's constant (8.617× 10^{-5} eV/K). Fig. (**11**) displays a plot of ln[(I_0/I) -1] *versus* 1/(kT). The thermal activation energy is 0.338 eV for α-Sr$_{1.99}$Eu$_{0.01}$P$_2$O$_7$ phosphor, which is obtained from the slope of linear fitting. The relatively high activation energy indicates that the phosphor could be applied for high-powered LED applications.

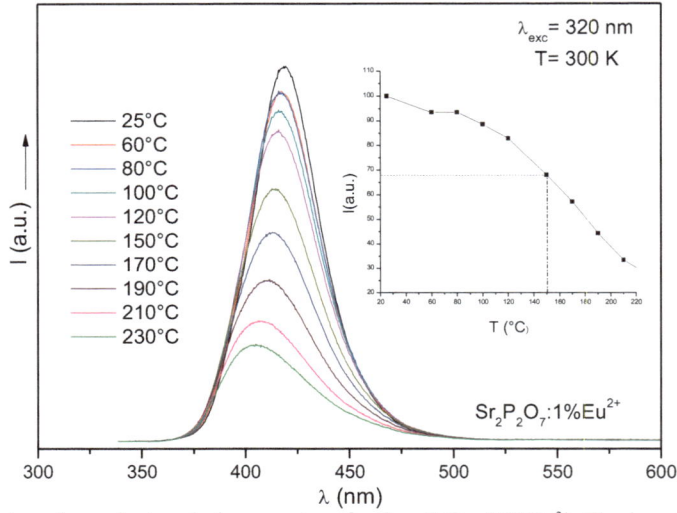

Fig. (10). Temperature dependent emission spectra of α-$Sr_{1.99}P_2O_7$: $0.01Eu^{2+}$. The insert shows temperature dependent relative emission intensity of α- $Sr_2P_2O_7$: $0.01Eu^{2+}$ sample.

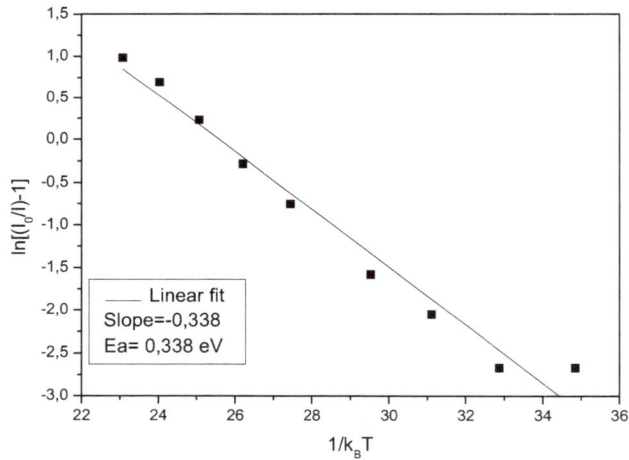

Fig. (11). A $\ln[(I_0/I)-1]$ *vs.* $1/k_BT$ activation energy graph for thermal quenching of α-$S_2P_2O_7$:1% Eu^{2+} phosphor.

CONCLUDING REMARKS

An intense blue α-$Sr_2P_2O_7$:Eu^{2+} phosphor was synthesized by a simple solid-state reaction method in a reducing atmosphere and its luminescence properties were up on VUV and UV excitation. The α-$Sr_2P_2O_7$:Eu^{2+} phosphor shows a wide and strong band located in the range of 250-400 nm matching with the n-UV LED chips. The decrease in emission intensities with increasing temperature as a result of a good thermal stability on the temperature quenching. The activation energy

Ea was calculated to be 0. 338 eV for α-Sr$_2$P$_2$O$_7$:Eu^{2+}. On the basis of our study, blue emitting phosphor α-Sr$_2$P$_2$O$_7$:Eu^{2+} is attractive as promising phosphors for application in White LEDs.

CONSENT FOR PUBLICATION

Not applicable.

CONFLICT OF INTEREST

The authors declare no conflict of interest, financial or otherwise.

ACKNOWLEGEMENTS

Declared none.

REFERENCES

[1]   M.H. Crawford, "LEDs for Solid-State Lighting: Performance Challenges and Recent Advances", *IEEE J. Sel. Top. Quantum Electron.,* vol. 15, pp. 1028-1040, 2009.
[http://dx.doi.org/10.1109/JSTQE.2009.2013476]

[2]   Q. Zhang, C.F. Wang, L.T. Ling, and S. Chen, "Fluorescent nanomaterial-derived white light-emitting diodes: What's going on", *J. Mater. Chem. C Mater. Opt. Electron. Devices,* vol. 2, pp. 4358-4373, 2014.
[http://dx.doi.org/10.1039/C4TC00048J]

[3]   J. Liu, Z.M. Zhang, Z.C. Wu, F.F. Wang, and Z. Li, "Study on luminescence and thermal stability of blue-emitting Sr$_5$(PO$_4$)$_3$F: Eu^{2+}phosphor for application in InGaN-based LEDs", *Mater. Sci. Eng. B,* vol. 221, pp. 10-16, 2017.
[http://dx.doi.org/10.1016/j.mseb.2017.03.014]

[4]   D. Böhnisch, J. Rosenboom, A. Garcia-Fuente, W. Urland, T. Jüstel, and F. Baur, "On the Blue Emitting Phosphor Na$_3$RbMg$_7$(PO$_4$)$_6$:Eu^{2+} Showing Ultra High Thermal Stability", *J. Mater. Chem. C Mater. Opt. Electron. Devices,* vol. 7, pp. 6012-6021, 2019. [http://dx.doi.org/10.1039/C9TC00482C]

[5]   D. Hou, C. Liu, X. Ding, X. Kuang, H. Liang, S. Sun, Y. Huang, and Y. Tao, "A high efficiency blue phosphor BaCa$_2$MgSi$_2$O$_8$:Eu^{2+} under VUV and UV excitation", *J. Mater. Chem. C Mater. Opt. Electron. Devices,* vol. 1, pp. 493-499, 2013.
[http://dx.doi.org/10.1039/C2TC00129B]

[6]   J. Zheng, L. Ying, Q. Cheng, Z. Guo, L. Cai, Y. Lu, and C. Chen, "Blue-emitting SrB$_2$O$_4$:Eu^{2+} phosphor with high color purity for near-UV white light-emitting diodes", *Mater. Res. Bull.,* vol. 64, pp. 51-54, 2015.
[http://dx.doi.org/10.1016/j.materresbull.2014.12.051]

[7]   "Kwon I.E, Park C.H, Hwang Y.J, Bae H.S, Yu B.Y, Pyun C.H, Hong G.Y. Phosphors for plasma display panels", *J. Alloys Compd.,* vol. 311, pp. 33-39, 2000.

[8]   A. Mbarek, G. Chadeyron, D. Boyer, D. Avignant, M. Fourati, D. Zambon, and R. Mahiou, "Vacuum ultraviolet excited luminescence properties of sol–gel derived GdP$_5$O$_{14}$:Eu^{3+} powders", *J. Lumin.,* vol. 145, pp. 335-339, 2014.
[http://dx.doi.org/10.1016/j.jlumin.2013.07.064]

[9]   J.L. Yuan, X.Y. Zeng, J.T. Zhao, Z.J. Zhang, H.H. Chen, and G.B. Zhang, "Rietveld refinement and photoluminescent properties of a new blue-emitting material: Eu^{2+} activated SrZnP$_2$O$_7$", *J. Solid State Chem.,* vol. 180, pp. 3310-3316, 2007.
[http://dx.doi.org/10.1016/j.jssc.2007.09.023]

[10]   M. Derbel, A. Mbarek, and M. Fourati, "Photoluminescence properties of $CdSrP_2O_7$:Eu^{2+} blue phosphor for white LED applications", *Optik (Stuttg.)*, vol. 127, pp. 5870-5875, 2016. [http://dx.doi.org/10.1016/j.ijleo.2016.04.020]

[11]   M. Derbel, and A. Mbarek, "Solid state synthesis and luminescent properties of bright blue-emitting $Ba_2P_2O_7$:Eu^{2+} phosphor", *SN Applied Sciences*, vol. 2, p. 562, 2020. [http://dx.doi.org/10.1007/s42452-020-2329-8]

[12]   S. Kolay, M. Basu, V. Sudarsan, and A.K. Tyagi, "Blue light emitting Eu doped Ca2P2O7 and $Ba_2P_2O_7$particles synthesized at low temperatures", *Solid State Sci.*, vol. 85, pp. 26-31, 2018. [http://dx.doi.org/10.1016/j.solidstatesciences.2018.09.007]

[13]   G. Blasse, and B.C. Gramaier, *Luminescent Materials*. Springer: Berlin, Heidelberg, 1994, p. 127. [http://dx.doi.org/10.1007/978-3-642-79017-1]

[14]   W.L. Wanmaker, and J.W. ter Vrugt, "Luminescence of calcium orthophosphates activated with divalent europium", *Philips Res. Pepts,* vol. 23, pp. 362-366, 1967.

[15]   S. Ye, Z-S. Liu, J.G. Wang, and X-P. Jing, "Luminescent properties of $Sr_2P_2O_7$:Eu,Mn phosphor under near UV excitation", *Mater. Res. Bull.*, vol. 43, pp. 1057-1065, 2008. [http://dx.doi.org/10.1016/j.materresbull.2007.07.039]

[16]   Y. Gao, Y. Cheng, F. Huang, H. Lin, J. Xu, and Y. Wang, "Sn^{2+}/Mn^{2+} codoped strontium phosphate ($Sr_2P_2O_7$) phosphor for high temperature optical thermometry", *J. Alloys Compd.*, vol. 735, pp. 1546-1552, 2018. [http://dx.doi.org/10.1016/j.jallcom.2017.11.243]

[17]   F. Le, L. Wang, W. Jia, D. Jia, and S. Bao, "Synthesis and photoluminescence of Eu^{2+} by co-doping Eu^{3+} and $Cl-$ in $Sr_2P_2O_7$ under air atmosphere", *J. Alloys Compd.*, vol. 512, pp. 323-327, 2012. [http://dx.doi.org/10.1016/j.jallcom.2011.09.088]

[18]   G. Ju, Y. Hu, L. Chen, X. Wang, and Z. Mu, "The influence of auxiliary codopants on persistent phosphor $Sr_2P_2O_7$:Eu^{2+},R^{3+} (R = Y, La, Ce, Gd, Tb and Lu)", *Mater. Res. Bull.*, vol. 48, pp. 4743-4748, 2013. [http://dx.doi.org/10.1016/j.materresbull.2013.08.011]

[19]   R. Pang, C. Li, L. Shi, and Q. Su, "A novel blue-emitting long-lasting proyphosphate phosphor $Sr_2P_2O_7$:Eu^{2+}, Y^{3+}", *J. Phys. Chem. Solids*, vol. 70, pp. 303-306, 2009. [http://dx.doi.org/10.1016/j.jpcs.2008.10.016]

[20]   D. S Thakare, S. K Omanwar, P. L Muthal, S. M Dhopte, V. K Kondawar, and S. V.Moharil,

[21]   L.O. Hagman, I. Jansson, and C. Magneli, "The crystal structure of α-$Sr_2P_2O_7$", *Acta Chem. Scand.*, vol. 22, pp. 1419-1429, 1968. [http://dx.doi.org/10.3891/acta.chem.scand.22-1419]

[22]   S. Carlino, M.J. Hudson, and W.J. Locke, "High-resolution solidstate magic-angle spinning nuclear magnetic resonance studies on the thermal decomposition of the layered antimony hydrogen phosphate, $HSb(PO_4)_2 \cdot 2H_2O$", *Solid State Ion.*, vol. 106, pp. 269-277, 1998. [http://dx.doi.org/10.1016/S0167-2738(97)00501-8]

[23]   S. Briche, D. Zambon, G. Chadeyron, D. Boyer, M. Dubois, and R. Mahiou, "Comparison of yttrium polyphosphate $Y(PO_3)_3$ preparedby sol–gel process and solid-state synthesis", *J. Sol-Gel Sci. Technol.*, vol. 55, pp. 41-51, 2010. [http://dx.doi.org/10.1007/s10971-010-2211-z]

[24]   A. Akrim, D. Zambon, J. Metin, and J.C. Cousseins, "Refinement of the $CsYP_2O_7$crystal-structure by the Rietveld method", *Eur. J. Solid State Inorg. Chem.*, vol. 30, pp. 483-495, 1993.

[25]   H.N. Ng, and C. Calvo, "The crystal structure of $KAlP_2O_7$", *Can. J. Chem.*, vol. 51, pp. 2613-2620, 1973. [http://dx.doi.org/10.1139/v73-395]

[26] V.P. Mahadevan Pillai, B.R. Thomas, V.U. Nayer, and K.H. Lii, "Infrared and Raman spectra of Cs$_2$VOP$_2$O$_7$ and single crystal Rb$_2$(VO)$_3$(P$_2$O$_7$)$_2$ Spectroc", *Acta Am.,* vol. 55, pp. 1809-1817, 1999. [http://dx.doi.org/10.1016/S1386-1425(99)00006-2]

[27] N. Khay, A. Ennaciri, and M. Harcharros, "Vibrational spectra of double diphosphates RbLnP$_2$O$_7$(Ln = Dy, Ho, Y, Er, Tm, Yb)", *Vib. Spectrosc.,* vol. 27, pp. 119-126, 2001. [http://dx.doi.org/10.1016/S0924-2031(01)00123-0]

[28] J. Tauc, R. Grigorovici, and A. Vancu, "Optical Properties and Electronic Structure of Amorphous Germanium", *Phys. Status Solidi, B Basic Res.,* vol. 15, pp. 627-637, 1966. [http://dx.doi.org/10.1002/pssb.19660150224]

[29] L.G. Van Uitert, "An empirical relation fitting the position in energy of the lower d-band edge for Eu^{2+} or Ce^{3+} in various compounds", *J. Lumin.,* vol. 29, pp. 1-9, 1984. [http://dx.doi.org/10.1016/0022-2313(84)90036-X]

[30] J. Sun, X. Zhang, and T. Li, "New Eu^{2+}-activated borophosphate phosphors Ba$_3$(ZnB$_5$O$_{10}$)PO$_4$ for near-ultraviolet-pumped white light-emitting diodes", *Mater. Lett.,* vol. 212, pp. 343-345, 2018. [http://dx.doi.org/10.1016/j.matlet.2017.10.125]

[31] Y. Zhang, Z. Xia, and W. Wu, "Preparation and luminescence properties of blue-emitting phosphor Ba$_2$Ca(PO$_4$)$_2$: Eu^{2+}", *J. Am. Ceram. Soc.,* vol. 96, pp. 1043-1046, 2013. [http://dx.doi.org/10.1111/jace.12258]

[32] M. Derbel, A. Mbarek, G. Chadeyron, M. Fourati, D. Zambon, and R. Mahiou, "Novel bluish white-emitting CdBaP2O7:Eu^{2+} phosphorfor near-UV white-emitting diodes", *J. Lumin.,* vol. 176, pp. 356-362, 2016. [http://dx.doi.org/10.1016/j.jlumin.2016.03.003]

[33] Z. Zhang, and Y. Wang, "Photoluminescence of Eu^{2+}-doped CaMgSi$_2$xO$_{6+2x}$ (1.00≤x≤1.20) phosphors in UV–VUV region", *J. Lumin.,* vol. 128, pp. 383-386, 2008. [http://dx.doi.org/10.1016/j.jlumin.2007.09.006]

[34] D.C. Tuan, R. Olazcuaga, F. Guillen, A. Garcia, B. Moine, and C. Fouassier, "Luminescent properties of Eu^{3+}-doped yttrium or gadolinium phosphates", *J Phys IV France,* vol. 123, pp. 259-263, 2005. [http://dx.doi.org/10.1051/jp4:2005123047]

[35] Z. Xia, R.S. Liu, K.W. Huang, and V. Drozd, "Ca$_2$Al$_3$O$_6$F:Eu^{2+}:a green emitting oxyfluoride phosphor for whitelight-emitting diodes", *J. Mater. Chem.,* vol. 22, pp. 15183-15189, 2012. [http://dx.doi.org/10.1039/c2jm32733c]

[36] L. Qin, P. Cai, C. Chen, J. Wang, S.I. Kim, Y. Huang, and H.J. Seo, "Optical performance of the Ba$_5$Al$_3$F$_{19}$: Eu^{2+} blue phosphors with high thermal stability", *J. Alloys Compd.,* vol. 738, pp. 372- 378, 2018. [http://dx.doi.org/10.1016/j.jallcom.2017.12.208]

[37] S. Shionoya, and W.M. Yen, *Phosphor Handbook.* CRC Press: New York, 1998.

[38] J.S. Kim, Y.H. Park, S.M. Kim, J.C. Choi, and H.L. Park, "Temperature-dependent emission spectra of M$_2$SiO$_4$:Eu^{2+} (M = Ca, Sr, Ba) phosphors for green and greenish white LEDs", *Solid State Commun.,* vol. 133, pp. 445-448, 2005. [http://dx.doi.org/10.1016/j.ssc.2004.12.002]

[39] G. Blasse, "Energy Transfer in Oxidic Phosphors", *Philips Res. Rep,* vol. 24, pp. 131-144, 1969.

[40] S. Bhushan, and M.V. Chukichev, "Temperature dependent studies of cathodoluminescence of green band of ZnO crystals", *J. Mater. Sci. Lett.,* vol. 7, pp. 319-321, 1988. [http://dx.doi.org/10.1007/BF01730729]

CHAPTER 7

Numerical Study of the Influence of Nano-fluid Type on Thermal Improvement in a Three Dimensional Mini Channel

Kamel Chadi[1,*], Nora Boultif[1], Nourredine Belghar[1], Aymen Mohamed Kethiri[1], Zied Driss[3] and **Belhi Guerira[2]**

[1] *Laboratory of Materials and Energy Engineering, University of Mohamed Khider Biskra, Algeria*

[2] *Laboratory of Mechanical Engineering, University of Mohamed Khider Biskra, Algeria*

[3] *Laboratory of Electromechanical Systems (LASEM), National School of Engineers of Sfax (ENIS), University of Sfax (US), B.P. 1173, Road Soukra km 3.5, 3038, Sfax, Tunisia*

Abstract: With the increasing development in the field of electronics, electronic devices have become smaller in size and more heat dissipating. This excessive heat leads to damage to the electronic components, and also their performance becomes bad. Therefore, the process of cooling them must be improved to increase their effectiveness in performance. For this purpose, we performed a numerical study to investigate the effect of different nanofluids on heat exchange in a silicon mini channel cooler for cooling electronic components. Three different types of nano-fluids were considered (TiO_2-H_2O, Ag-H_2O, and SWCNT-H_2O). In this study, the volumetric fraction of nano-particles is taken to be 2%, the Reynolds number (Re) is varied between 100 and 700, and the flow regime is assumed to be stationary. The ANSYS Fluent 17.1 commercial software is used as a calculation tool to solve the governing equations, which depend on the finite volume method (FVM) in its solution. The relaxation of decreasing factors used in this study is 0.7 and 0.3 to maintain momentum and pressure, respectively. The residual values of the continuity equation and velocity components are in the range of 10^{-5} and 10^{-6}, respectively, and the second-order upwind scheme has been used. The results obtained show that the maximum temperature of the electronic component decreases with the increase in the Reynolds number. The reduction in the temperature of the electronic component is more noticeable for the TiO_2-water and SWCNT water nano-fluid. As we found that the values of the coefficient of heat exchange between the channel walls and the nano-fluid that contains the single-walled carbon nanotubes nanoparticles are the highest compared to the nanoparticles that do not contain carbon in their composition, therefore, this condition can be considered the best in heat transfer. Therefore, it is recommended that nanofluids containing nanoparticles SWCNT for cooling high-temperature electronic components should be used.

[*] **Corresponding author Chadi Kamel:** Laboratory of Materials and Energy Engineering, University of Mohamed Khider Biskra, Algeria; E-mail: chadikamel_dz@yahoo.fr

Zied Driss (Ed.)

Keywords: ANSYS Fluent, Nano-Fluid Type, Silicon Mini Channels, SWCNT Nano-Particles, Thermal Exchange.

INTRODUCTION

Temperature is an important factor in the functioning of electronic components. This temperature can be sometimes sufficiently high, which decreases the lifespan of the electronic components. To improve this problem regarding temperature, it is necessary to choose the appropriate liquid for cooling electronic components. Among the researches in the domain of cooling electronic component, Nguyen *et al.*'s study involved an experimental investigation for the enhancement of heat transfer coefficient of Al_2O_3–H_2O nano-fluid [1]. They concluded that an increase of particle volume fraction in water has produced a clear decrease of the electronic component temperature and has lead to a significant improvement of heat exchange coefficient. Furthermore, their experimental results showed that nano liquid with 36 nm particle size gives higher thermal exchange coefficients than the ones with 47 nm particle diameter. Khaleduzzaman *et al.* presented an analytical study of the thermal performance improvement of three nano-liquids (Al_2O_3-H_2O, CuO-H_2O, SiC- H_2O) for a copper rectangular micro-channel for electronic component cooling [2]. They concluded that CuO-water nano-fluid is the most suitable for cooling electronics among three nano-fluids.

Gholinia *et al.* conducted a physical study of the aspects of the steady boundary layer of CNTs ($C_2H_6O_2$–H_2O) hybrid base nanoparticles over a porous expanding cylinder under the influence of magnetic force [3]. They also studied the effect of different nanoparticles on the Reynolds number and Nusselt number as they concluded that the cooling/heating process depends on the selection of nanoparticles types.

Also, Moosavi and others analyzed physical aspects of magnetic convection and heat transfer of a TiO_2–Al_2O_3/$C_2H_6O_2$–H_2O hybrid base nano-fluid inside a porous container under the influence of Lorentz forces and the form factor in this paper; the effect of different parameters, such as Hartmann's number, radiation coefficient, volume fraction and buoyancy forces on the temperature gradient and Nusselt number was studied [4]. They also concluded that nanoparticles with a lamina shape have a greater effect on increasing the average Nusselt number compared to the shape factor.

In the same context, we also found that Gholinia and the others analyzed the flow stagnation point characteristics of the three-dimensional hybrid nanofluids passing through a circular cylinder with a sinusoidal radius. In this study, they used the SWCNT and MWCNT nanoparticles [5].

They also used engine oil, water, and $C_2H_6O_2$ oil as the base liquid. Besides, they examined the Prandtl number, fraction size of carbon nanotube, and thermal slip parameter on the heat transfer rate of various carbon nanotubes. Among the most important results, they found out that the cooling process could be increased using a smaller thermal slip parameter.

Moreover, Ghadikolaei and Gholinia studied the effect of thermal radiation, slip factors, and shape, by using a nanofluid containing GO-MoS$_2$ particles and water-ethylene glycol ($C_2H_6O_2$) (50:50 Vol%) [6]. They also studied the parameters, such as the shapes of the nanoparticles (cylinders, platelets, and bricks), the suction/injection parameters on the temperature profiles, and the speed. Their results suggest that an increase in the form factor leads to an increase in the heat transfer rate and the temperature profile. They also concluded that the strong hydrogen bonding of the hybrid nanofluid (GO-MoS$_2$) leads to an increase in heat exchange.

Shakhaoath *et al.* studied the unstable flow of the boundary layer of the convection of a nanofluid along with an extended sheet with thermal radiation in the presence of a magnetic field [7]. They also studied the effects of speed, temperature, and concentration on the sheet, highlighting that Brownian motion had an effect on the stability of boundary layer growth. They also noted that the use of the magnetic field could control the flow properties, and the thermal boundary layer thickness increased due to the increase in radiation.

Arifuzzaman *et al.* modeled the unstable natural convection of a high-velocity MHD liquid flow near a semi-infinite plate, moving the porous plate with thermal radiation [8]. The flow is created by a porous plate that oscillates vertically. They also studied the velocity, temperature, and concentration in the boundary layer region. Among the most important results, it was found that as the cross-diffusion, magnetic parameter, heat sink, radiation absorption, and thermal radiation are increased, the temperature increases. They also observed that while the cross-diffusion, Prandtl number, and magnetic parameter decrease, the Nusselt number decreases.

Also, Gholinia *et al.* studied the physics aspects of magnetic and heat exchange of a hybrid base nano-fluid in a porous medium under the impact of shape, Lorentz force, and thermal radiation [9]. In this study, they used the finite element method to solve the governing equations and also used a mixture of copper oxide nanoparticles in different shapes in ethylene glycol 50%-water 50% (similarly for Fe_3O_4). In their paper, they talked about the effect of various parameters, such as Hartmann (Ha) number, volume fraction, radiation parameter, and buoyancy force, temperature gradient, and Nusselt number. Their results showed that Fe_3O_4

nanoparticles have a smaller temperature gradient than the particles containing copper oxide. They concluded that the blade-shaped nanoparticles have a greater effect on increasing the Nusselt number.

On the other hand, Ghobadi and Hassankolaei [10] reviewed the magnetic flux of a carreau nanoliquid on the radiative expansion plate. In this study, they applied the Brownian motion to the model nanoparticles, where the effect of different factors on the change in velocity, concentration, temperature, and Nusselt number was studied in two cases (the shear-thinning fluid and the shear thickening fluid). Their results showed that the temperature profile has direct relationships with the magnetic field and thermal radiation.

Also, Ghobadi and Hassankolaei [11] studied and analyzed the heat transfer and magnetohydrodynamic stagnation point flow of the hybrid nanoparticles. In this study, they used the hybrid nanofluid Al_2O_3-TiO_2/H_2O to improve the heat transfer coefficient. They also studied the effect of some parameters, such as the shape factor of nanoparticles and lamina, on the velocity profile, temperature, and the magnetic field. One of the most important conclusions they reached is that the shape of the lamina nanoparticles has a greater effect on the Nusselt number compared to the hexagonal-shaped nanoparticles. They also found out that an increase in the shape factor, solar radiation, and the use of hybrid nanoparticles lead to an increase in the temperature profile.

The main thrust of this article is the use of various nanoparticles, especially single-walled carbon nanotubes nanoparticles, to improve thermal exchange in a finned minichannel in order to increase the cooling efficiency of electronic components.

In this paper, ANSYS FLUENT is used to solve the governing equations, which depend on the finite volume method (FVM). The SIMPLE algorithm is also used (semi-implicit method for pressure-related equation) to implement pressure and velocity coupling.

STUDIED GEOMETRIES

Fig. (1) shows the geometry of the studied mini channel cooler using fluent industrial software. The dimensions of the cooler are in the order of 40 x 52 mm^2 with a thickness of 6 mm. This cooler consists of 13 channels made of silicon, and we placed a heat insulator at the bottom of the cooler, and at the top of the cooler, there is an electronic component represented in an electronic chip with a constant power of 150 watts with thermal insulation on all the outer faces of the cooler. Each channel has two inclined fins at an angle of 45°, the mission of which is to increase the efficiency of heat exchange between the walls of the mini channel and the coolant. The inlet temperature of the nano-fluids (TiO_2-water, Ag-water, and SWCNT-water) in the cooler is 293K.

Due to symmetry, and in order to reduce the grid size and the computational time, only half of the mini channels have been modeled.

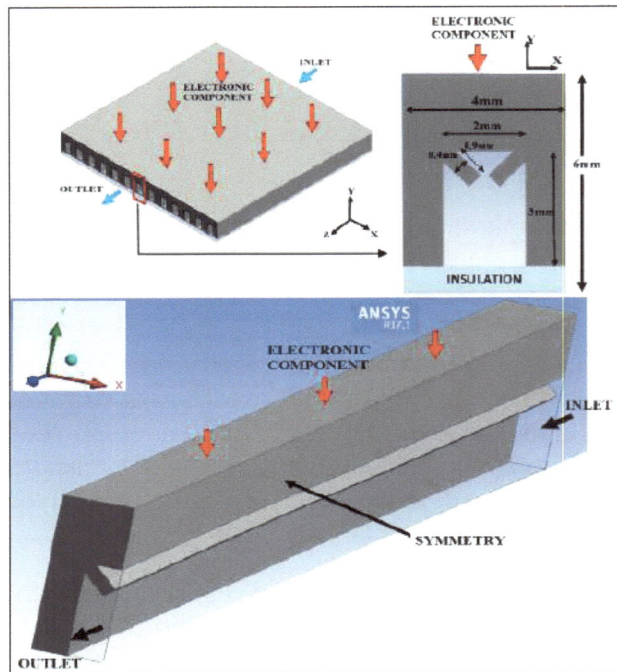

Fig. (1). Schematic representation of the studied mini channels cooler and computational domain of mini-channel heat sink.

MATHEMATICAL FORMULATION

In this research, we assumed that the flow is laminar flow and stationary. The three nano-fluids (TiO_2-water, Ag-water, and SWCNT-water) are supposed to be incompressible, and the thermo-physical properties of these nano-fluids are constant. The thermal exchange by radiation is considered negligible.

The boundary conditions are:

- At the outlet of the mini channels, the pressure is zéro (P = 0)
- The velocity components (u, v, w) of the nanofluid at the level of the channel wall are equal to zero, and no-slip boundary conditions are applied to all mini channel walls.
- The impact of body force and viscosity dissipation is neglected

- At the inlet flow, the speed and the temperature of nano liquid are constant
- At the nanofluid/solid interface, the continuity of the heat flux at the interface between the solid of the mini channel and the nano liquid is implemented using:

$$k_s \frac{\partial T}{\partial n}\bigg|_{wall} = k_f \frac{\partial T}{\partial n}\bigg|_{wall} \tag{1}$$

The formulas for the calculation of the thermophysical properties of the nano-liquids utilized in this study are written as follows:

- The effective thermal conductivity of the TiO_2-water, Ag-water, and SWCNT-water nano-fluids are approximated by Maxwell-Granetts, as mentioned-below [12]:

$$k_{nf} = \frac{k_S + 2k_f - 2\varphi(k_f - k_S)}{k_S + 2k_f + \varphi(k_f - k_S)} k_f \tag{2}$$

Where k_s, k_f and φ represents the thermal conductivity of the solid nanoparticles, the thermal conductivity of fluid (water), and volume fraction, respectively.

- The dynamic viscosity (μ_{nf}) is approximated by the equation given below [12]:

$$\mu_{nf} = \frac{\mu_f}{(1-\varphi)^{2,5}} \tag{3}$$

Where μ_f is the dynamic viscosity of coolant.

- The density of the nano liquids is given as follows [13]:

$$\rho_{nf} = (1-\varphi)\rho_f + \varphi\rho_s \tag{4}$$

Where ρ_f and ρ_s are the density of the base fluid and the solid nanoparticles

- The heat capacitance of the nano liquids is given as [13]:

$$(\rho C_P)_{nf} = (1-\varphi)(\rho C_P)_f + \varphi(\rho C_P)_s \tag{5}$$

Where C_p is the specific heat of the liquid.

The thermophysical properties of base fluid and Ag, TiO_2, and SWCNT nanoparticles are presented in Table **1**.

Table 1. The thermo-physical properties of H_2O and TiO_2, Ag, and SWCNT nanoparticles.

The Thermophysical Properties	Water (Base Fluid) [14]	TiO_2 Nanoparticles [14]	Ag Nanoparticles [14]	SWCNT Nanoparticles [15]
Density (kg/m³)	998.2	4250	10.500	2600
Specific heat (Jkg⁻¹ K⁻¹)	4182	686.2	235	425
Thermal Conductivity (Wm⁻¹K⁻¹)	0.613	8.9538	429	6600
Dynamic Viscosity (kg/m.s)	0.001003	/	/	/

The governing equations are given in terms of the continuity equation, momentum equation, energy conservation equation, and solid equation [16]. The governing equations for carbon nanotubes have been mentioned in a study [17].

• The continuity equation is written as follows:

$$\frac{\partial u}{\partial x} + \frac{\partial v}{\partial y} + \frac{\partial w}{\partial z} = 0 \qquad (6)$$

Where *u, v,* and *w* are the speed components in the *x*-axis, *y*-axis, *z*-axis, respectively.

• The momentum equations along the *x*-axis, the *y*-axis, and the *z*-axis are written as follows:

x-axis:

$$u\frac{\partial u}{\partial x} + v\frac{\partial u}{\partial y} + w\frac{\partial u}{\partial z} = \frac{1}{\rho_{nf}}\left[-\frac{\partial P}{\partial x} + \mu_{nf}\left(\frac{\partial^2 u}{\partial x^2} + \frac{\partial^2 u}{\partial y^2} + \frac{\partial^2 u}{\partial z^2}\right)\right] \qquad (7)$$

y-axis:

$$u\frac{\partial v}{\partial x} + v\frac{\partial v}{\partial y} + w\frac{\partial v}{\partial z} = \frac{1}{\rho_{nf}}\left[-\frac{\partial P}{\partial y} + \mu_{nf}\left(\frac{\partial^2 v}{\partial x^2} + \frac{\partial^2 v}{\partial y^2} + \frac{\partial^2 v}{\partial z^2}\right)\right] \qquad (8)$$

z-axis:

$$u\frac{\partial w}{\partial x} + v\frac{\partial w}{\partial y} + w\frac{\partial w}{\partial z} = \frac{1}{\rho_{nf}}\left[-\frac{\partial P}{\partial z} + \mu_{nf}\left(\frac{\partial^2 w}{\partial x^2} + \frac{\partial^2 w}{\partial y^2} + \frac{\partial^2 w}{\partial z^2} \right) \right] \qquad (9)$$

where ρ_{nf}, μ_{nf} and P are the density, dynamic viscosity and the pressure of the nano fluid, respectively.

• The energy equation is written as follows:

$$u\frac{\partial T}{\partial x} + v\frac{\partial T}{\partial y} + w\frac{\partial T}{\partial z} = \alpha_{nf}\left(\frac{\partial^2 T}{\partial x^2} + \frac{\partial^2 T}{\partial y^2} + \frac{\partial^2 T}{\partial z^2} \right) \qquad (10)$$

Where T is the temperature of the nanofluid, and αnf is the thermal diffusivity of the nano fluid.

• The heat conduction through the solid wall can be calculated through the following equation:

$$\frac{\partial^2 T_S}{\partial x^2} + \frac{\partial^2 T_S}{\partial y^2} + \frac{\partial^2 T_S}{\partial z^2} = 0 \qquad (11)$$

RESULTS AND INTERPRETATIONS

In this work, a hexahedron mesh is chosen (structured mesh) in the three directions (x, y and z); after the simulation calculations are converged, we tested the mesh independently. (Fig. **2**) shows a meshing view in three dimensions (3D). Meshing details include the number of nodes, which are equal to 459546.

In the simulation, we obtained a convergence for the studied model. The nodes used for meshing of the physical domain are affecting the results. (Fig. **3**) shows the number of nodes used for the analysis of the mini channel heat sink with the result of the temperature of the electronic component. According to this figure, the results concluded from 900,000 nodes is that the solution is independent of the mesh.

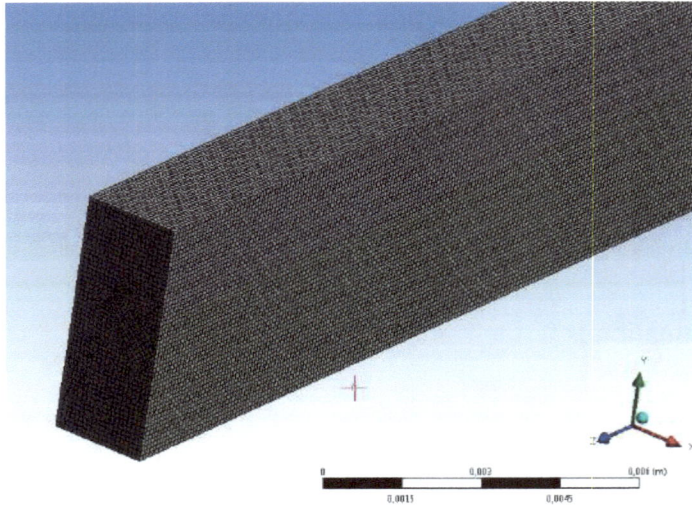

Fig. (2).   Meshing of minichannel.

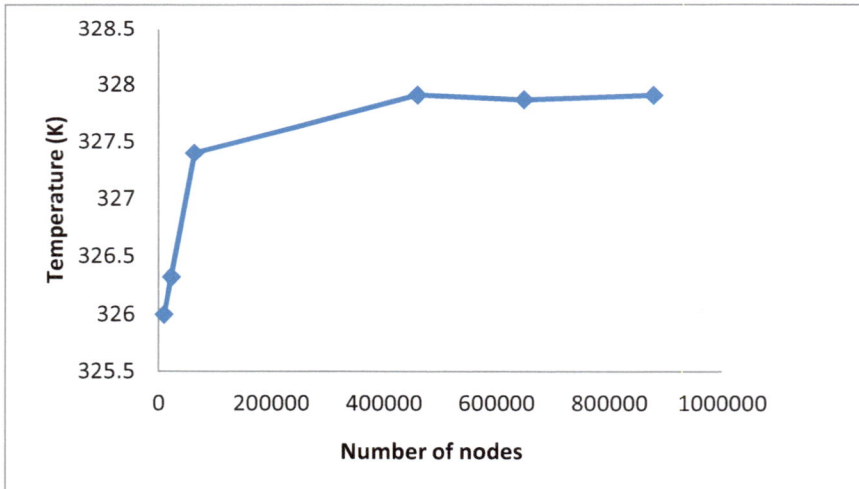

Fig. (3).   Grid independence examination.

Fig. (**4**) shows the evolution of the maximum temperature value of the electronic component with respect to the Reynolds number. The temperature gradually decreases for the three types of nano fluids as the Reynolds number increases. We note that in all studied nanofluids, the temperature value of the electronic component is high at the lowest values of the Reynolds number, and the temperature is low at the highest Reynolds number. We also note that the type of nanoparticles greatly affects the maximum temperature of the electronic component, as the Titanium dioxide nanoparticles contribute better to lower temperature of the electronic component compared to the Ag nanoparticles, whereas, we found out that single-walled carbon nanotubes nanoparticles are better compared to Ag and titanium dioxide nanoparticles, and this is due to the good physical properties that characterize the SWCNT nanoparticles.

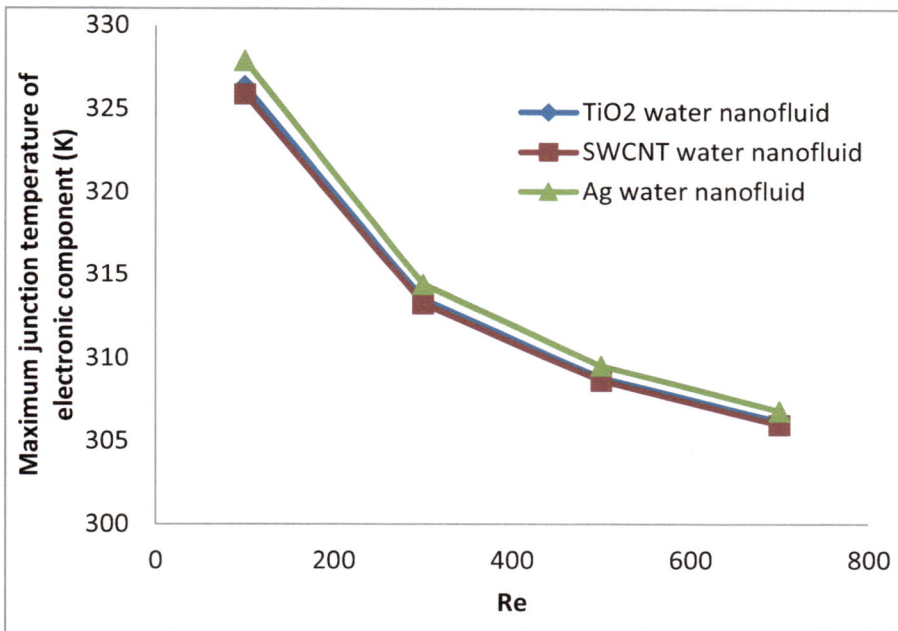

Fig. (4). The variation of the temperature of the electronic component according to the number of Reynolds.

Fig. (**5**) shows the effect of Reynolds number on the temperature values at the outlet of the mini channel for the volume fraction of Ag nanoparticles equal to 2%. According to the results of (Fig. **4**), we note that the temperature of the nanofluid Ag-Water near the walls of the mini channel at the Reynolds number equal to 700 is low; on the other hand, when the Reynolds number = 100, the temperature values near the walls of the mini channel at the outlet flow are high. The low temperature was achieved when the flow velocity of the nano liquid near the mini channel walls was increased. On the other hand, we notice that the fins significantly affect the temperature values of the nanofluid, and it can be said that the fins inside the mini channel are effective in cooling when the liquid flow velocity is highest, *i.e*, when the Reynolds number equals 700. We find in Fig.

(**6**) a comparison among the three nanoscale fluids used in terms of the average heat transfer coefficient according to the variable Reynolds number values from 100 to 700. According to the results of (Fig. **6**), we conclude that when the fluid flow velocity is increased, the nanoparticles are in a rapid random transfer state within the base fluid, which contributes to an increase in heat exchange between fluid and mini channel walls. However, in this research, the results show that nanofluid containing SWCNT nanoparticles is the best in terms of the heat exchange rate, therefore, it can be considered suitable in the process of cooling electronic components.

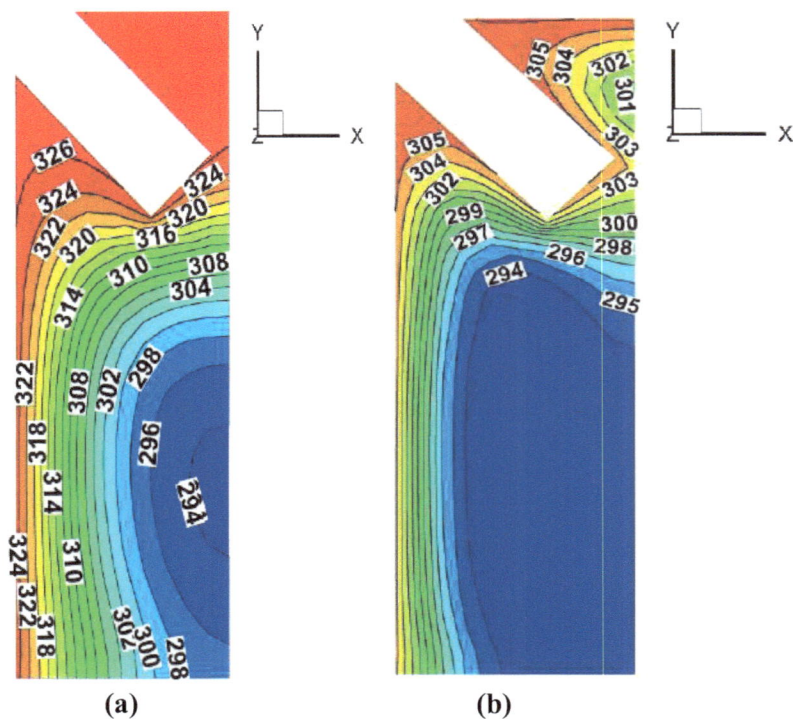

(a)   (b)

Fig. (5). The distribution of the temperature in the outlet of mini channel for (a) Re =100 and (b) Re = 700.

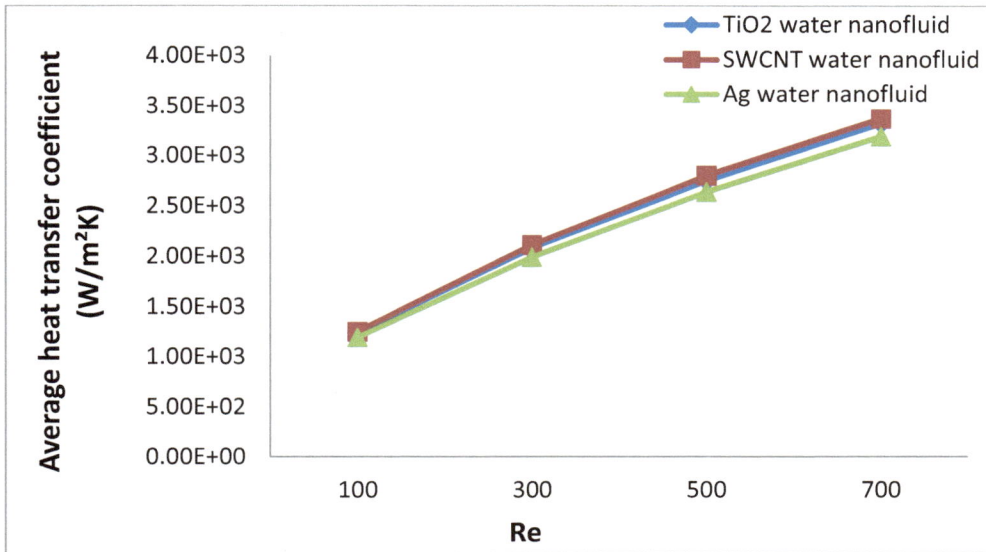

Fig. (6). The average heat transfer coefficient calculated as a function of the Reynolds number at a volume fraction of 2%.

CONCLUSION

In this research, a numerical study is conducted to find out the effect of the nature of nanofluids (Ag-water, TiO_2-water, and SWCNT-water) on the temperature of the electronic component and the heat exchange of the mini channel with fins. The ANSYS Fluent 17.1 has been used to solve the governing equations, which depend on the finite volume method (FVM). The residual values of the continuity equation and velocity components are in the range of 10^{-5} and 10^{-6}, respectively; the second-order upwind scheme has been used. The results obtained confirm that the temperature of the electronic component decreases with the increase in the Reynolds number and the flow velocity. Furthermore, the use of SWCNT-water and TiO_2-water nanofluids helps reduce the temperature of the electronic component compared to the nanofluid containing Ag nanoparticles; the use of SWCNT-water nanofluid gives better heat transfer coefficients compared to Ag-water and TiO_2-water. In addition, as a future study, the properties of cooling liquids can be increased and improved by using hybrid nanofluids with acceptable conductivity.

CONSENT FOR PUBLICATION

Not Applicable.

CONFLICT OF INTEREST

The authors declare no conflict of interest, financial or otherwise.

ACKNOWLEDGEMENTS

Declared none.

REFERENCES

[1]　C.T. Nguyen, G. Roy, C. Gauthier, and N. Galanis, "Heat transfer enhancement using Al2O3-water nano-fluid for an electronic liquid cooling system", *Appl. Therm. Eng.,* vol. 27, pp. 1501-1506, 2007. [http://dx.doi.org/10.1016/j.applthermaleng.2006.09.028]

[2]　S.S. Khaleduzzaman, R. Saidur, J. Selvaraj, I.M. Mahbubul, M.R. Sohel, and I.M. Shahrul, "Nano-fluids for Thermal Performance Improvement in Cooling of Electronic Device", *Adv. Mat. Res.,* vol. 832, pp. 218-223, 2014.

[3]　M. Gholinia, M. Armin, A.A. Ranjbar, and D.D. Ganji, "Numerical thermal study on CNTs/C2H6O2-H2O hybrid base nanofluid upon a porous stretching cylinder under impact of magnetic source", *Case Stud. Therm. Eng.,* vol. 14, 2019.100490 [http://dx.doi.org/10.1016/j.csite.2019.100490]

[4]　S A H Kiaeian Moosavi, M Pourfallah, S Gholinia, D Ganji D, and M Gholinia, A numerical treatment of the TiO2/C2H6O2–H2O hybrid base nanofluid inside a porous cavity under the impact of shape factor in MHD flow., *International Journal of Ambient Energy,* 2019.

[5]　M Gholinia, Kh Hosseinzadeh, and D.D. Ganji, "Investigation of different base fluids suspend by CNTs hybrid nanoparticle over a vertical circular cylinder with sinusoidal radius", *Case Studies in Thermal Engineering,* vol. 21, pp. 100-666, 2020.

[6]　S.S. Ghadikolaei, and M. Gholinia, "3D mixed convection MHD flow of GO-MoS2 hybrid nanoparticles in H2O–(CH2OH)2 hybrid base fluid under the effect of H2 bond", *Int. Commun. Heat Mass Transf.,* vol. 110, 2020.104371 [http://dx.doi.org/10.1016/j.icheatmasstransfer.2019.104371]

[7]　K. Md Shakhaoath, K. Ifsana, and E. Lasker, "A and Ariful I. Unsteady MHD free convection boundary-layer flow of a nanofluid along a stretching sheet with thermal radiation and viscous dissipation effects", *Int. Nano Lett.,* vol. 2, p. 24, 2012. [http://dx.doi.org/10.1186/2228-5326-2-24]

[8]　S.M Arifuzzaman, M.S Khan, M.F.U Mehedi, B.M.J Rana, and S.F. Ahmmed, "Chemically reactive and naturally convective high speed MHD fluid flow through an oscillatory vertical porous plate with heat and radiation absorption effect", *Engineering Science and Technology, an International Journal ,* vol. 21, pp. 215-228, 2018.

[9]　M. Gholinia, S.A.H. Kiaeian Moosavi, S. Gholinia, and D.D. Ganji, "Numerical simulation of nanoparticle shape and thermal ray on a CuO/C2H6O2–H2O hybrid base nanofluid inside a porous enclosure using Darcy's law", *Heat Transf. Asian Res.,* vol. 48, pp. 3278-3294, 2019. [http://dx.doi.org/10.1002/htj.21541]

[10]　A.H. Ghobadi, and M.G. Hassankolaei, "Numerical treatment of magneto Carreau nanofluid over a stretching sheet considering Joule heating impact and nonlinear thermal ray", *Heat Transf. Asian Res.,* vol. 48, pp. 4133-4151, 2019. [http://dx.doi.org/10.1002/htj.21585]

[11]　A.H. Ghobadi, and M.G. Hassankolaei, "Numerical approach for MHD Al2O3-TiO2/H2O hybrid nanofluids over a stretching cylinder under the impact of shape factor", *Heat Transf. Asian Res.,* vol. 48, pp. 4262-4282, 2019.

[http://dx.doi.org/10.1002/htj.21591]

[12]   S. Fohanno, G. Polidori, and C. Popa, *Nanofluides et transfert de chaleur par convection naturelle.* Université de reims champagne-Ardenne: France, 2012.

[13]   K. Khanafer, K. Vafai, and M. Lightstone, "Buoyancy-driven heat transfer enhancement in a two dimensional enclosure utilizing nanofluids", *Int. J. Heat Mass Transf.,* vol. 46, pp. 3639-3653, 2003. [http://dx.doi.org/10.1016/S0017-9310(03)00156-X]

[14]   H.A. Mohammed, P. Gunnasegaran, and N.H. Shuaib, "The impact of various nanofluid types on triangular microchannels heat sink cooling performance", *Int. Commun. Heat Mass Transf.,* vol. 38, pp. 767-773, 2011. [http://dx.doi.org/10.1016/j.icheatmasstransfer.2011.03.024]

[15]   A.K. Waqar, C. Richard, and H. Rizwan Ul, "Heat Transfer Analysis of MHD Water Functionalized Carbon Nanotube Flow over a Static/Moving Wedge", In: *Journal of Nanomaterials,* 2015, p. 13.

[16]   M.K. Moraveji, R.M. Ardehali, and A. Ijam, "CFD investigation of nano-fluid effects (cooling performance and pressure drop) in mini-channel heat sink", *Int. Commun. Heat Mass Transf.,* vol. 40, pp. 58-66, 2013. [http://dx.doi.org/10.1016/j.icheatmasstransfer.2012.10.021]

[17]   A.A. Abbasian Arani, O.A. Akbari, M. Reza Safaei, and A. Marzban, "A.A.A. Alrashed A, Reza Ahmadi G, Khang Nguyen T. Heat transfer improvement of water/single-wall carbon nanotubes (SWCNT) nanofluid in a novel design of a truncated double-layered microchannel heat sink", *Int. J. Heat Mass Transf.,* vol. 113, pp. 780-795, 2017. [http://dx.doi.org/10.1016/j.ijheatmasstransfer.2017.05.089]

CHAPTER 8

Electrochemical Studies and Characterization of Zn-Mn Coatings deposited in the presence of Novel Organic Additives

Nouha Loukil[1,*] and **Mongi Feki**[2]

[1] *Laboratory of Material Engineering and Environment, ENIS-Tunisia, University of Sfax, Sfax, Tunisia*

[2] *Laboratory of Material Engineering and Environment, ENIS-Tunisia, University of Sfax, Sfax, Tunisia*

Abstract: A novel additive based on alkylphenol ethoxylate sulphite was investigated in Zn-Mn electrodeposition on steel from a chloride bath. Electrochemical study via cyclic voltammetry showed that the tested additive increases the over-potential of the Zn deposition, resulting from strong adsorption of molecules additives on the cathode surface. Thus, Mn-rich alloy containing 16.3% of Mn is successfully co-deposited. The morphology and crystallographic structure of Zn-Mn co-deposits were analyzed using Scanning Electron Microscopy (SEM) and X-Ray Diffraction (XRD), respectively. SEM micrographs showed that Zn-16.3% Mn alloy obtained in the presence of the tested additive displays hexagonal pyramid morphology. XRD analysis exhibited that Zn-16.3% Mn alloy is monophasic with hexagonal close-packed ε-Zn-Mn phase.

Keywords: Additive, Electrodeposition, Morphology, Structure, Zn-Mn alloy.

INTRODUCTION

Numerous zinc-based coatings Zn-X (X= Fe, Co, Ni, Mn...) have been electrodeposited on steel to improve the corrosion resistance of conventional pure zinc [1]. Currently, fundamental researches have been made to assess the corrosion resistance of these binary alloys into automotive applications [2 - 5]. Among Zn binary alloys, there is a growing interest in Zn-Mn coatings owing to the highest anticorrosive properties in the saline environment [4 - 7]. Indeed, Zn-Mn alloys show sacrificial corrosion protection for both Zn and steel components. Boshkov *et al.* reported that the alloying element Mn has been proposed due to its dual protective role [8, 9]. Mn is anodic to Zn and primarily dissolves [8, 9].

---

[*] **Corresponding author Nouha Loukil:** Laboratory of Material Engineering and Environment, ENIS-Tunisia, University of Sfax, Sfax, Tunisia; E-mail: nloukil87@gmail.com

Accordingly, pH locally increases and promotes the formation of $Zn_5(OH)_8Cl_2.H_2O$ **(ZHC)** and $Zn_5(OH)_6SO_4xH_2O$ **(ZHS)** protective layers in chloride and sulfate aggressive media, respectively [8, 9]. These protective layers are thick, compact, uniform, and with low solubility [8 - 13]. This is recognised by the homogeneous distribution of Mn in the coatings [14].

Several existing reports investigated the effect of the Mn content on the corrosion behavior of Zn-Mn coatings. Zn-11% Mn shows the highest anticorrosive resistance [10, 11]. Muller *et al.* proved that 10 to 40% of Mn substantially affords better corrosion resistance [11]. It is well documented that low Mn content ranging from 0.3 to 2.5% provides suitable mechanical properties for automotive applications and good plastic deformability [7], weldability and paintability. However, Zn-Mn co-deposition from aqueous solutions is a complex process in view of the large gap between the deposition potentials of Zn ($E°$ (Zn^{2+}/Zn) = -0.76 V/HSE and Mn ($E°$ (Mn^{2+}/Mn) = -1.18 V/HSE). These two potentials are more negative than that of hydrogen evolution, which is a concurrent reaction.

According to Brenner [15], Zn and Mn co-deposition is a normal co-deposition, implying that Mn-rich coatings require high current densities to be electrodeposited [1, 13, 14, 16]. This occurrence leads to **(i)** low current efficiency [13, 14, 16] and **(ii)** burned and non-adherent Zn-Mn deposits [13, 14, 16]. These drawbacks result from the intensive hydrogen evolution reaction at high cathodic potentials required to reduce Mn^{2+} ions.

Zn-Mn coatings deposited from aqueous solution have comprehensively been reported in the literature. Main electrolytic baths being used are, chloride-based [1, 6, 17 - 19], pyrophosphate-based [6, 18] or citrate-based electrolytes [18, 20].

One way to partially overcome the drawbacks of Zn-Mn co-deposition is the use of additives. Using a strong organic complexing agent for Zn^{2+} ions is the common way to reduce the gap between Zn and Mn deposition potentials [2, 6, 7, 11, 13, 15]. This is ascribed to the high relative formation constant of the complex formed with Zn^{2+} ions [11]. Meanwhile, they must be avoided due to environmental problems related to effluent treatment.

Few alternative organic additives are reported in the literature, which generally fall into two classes: the carrier additive and the brightener. Commonly, the carrier additive permits the **(i)** minimization of the potential gap between Zn and Mn co-deposition and/or hydrogen evolution [1, 17] and **(ii)** the grain refinement [17, 19], while the brightener additive has a complementary effect to obtain bright deposits [17, 21].

The present study deals with the development of a suitable formulation for Zn-Mn bath, including a proprietary additive. The additive formulation includes two parts: a carrier and a brightener, which are alkylphenol ethoxylate sulphite and Benzenlideneacetone (BDA), respectively. This formulation has not been explored before. The aim of the present work was to study the influence of the retained additive on Zn^{2+} and Mn^{2+} ions reduction from chloride acidic bath through cyclic voltammetry (CV). Morphological and structural characterizations of Zn-Mn coatings will also be discussed.

EXPERIMENTAL DETAILS

The electrodeposition tests were performed using an acidic chloride electrolyte. Blank experiments were carried out in three basic electrolytes S_{Zn}, S_{Mn}, and S_{Zn-Mn}. The compositions of all electrolytic solutions are shown in Table **1**.

Table 1. Composition of electrolytic solutions.

	S_{Zn}	S_{Mn}	S_{Zn-Mn}
KCl	3.2	3.2	3.2
H_3BO_3	0.4	0.4	0.4
$ZnCl_2$	0.4	-	0.4
$MnCl_2$	-	0.4	0.4

The tested additive is composed of two parts, namely carrier, and brightener. The formulation of this additive is complex and the base part is an alkylphenol ethoxylated sulphite sodium salt, which belongs to a non-ionic surfactant. An alkylphenol ethoxylate usually consists of a branched-chain nonylphenol reacted with an ethylene oxide [22]. The brightener used for this study is Benzenlideneacetone (**BDA).**

All reagents used were of analytical grade. The electrolytic pH was regularly adjusted to 5 using dilute hydrochloric acid (HCl) and potassium hydroxide (KOH) solution. The unstirred plating baths were operated at room temperature (25 °C).

Electrochemical experiments were carried out in a conventional three-electrode cell using an Autolab PGSTAT302N controlled by NOVA software, allowing data acquisition. All potentials were referred to Ag/AgCl reference electrode. A pure Zn plate was used as a counter electrode for both electrochemical studies and bulk electrolysis.

For electrochemical studies, the working electrode was a steel disk of 1 cm diameter. The specimens were cold-covered in epoxy resin. Prior to each electrodeposition, steel surface preparation is as follows: mechanical polishing, degreasing the surface in alcohol, and then pickling in a chloride medium (HCl) to remove the oxide layer on the surface. It is important to note that hydrochloric acid (HCl) fastly pickles, minimizing then the base metal loss.

For bulk electrolysis, the steel substrate (30×40 mm^2) placed as cathode and a soluble Zn anode were connected to a digital dc power supply providing current and voltage. The bulk electrolysis tests were performed for 30 min under a constant current density of 15 mA/cm^2.

Cyclic Voltammetry

Voltammetric measurements at the motionless electrode were carried out to better understand the electrochemical behavior of Zn and Mn separately in the presence and absence of the additive.

The potential domain was scanned from -0.6 V to the cathodic direction up to - 2 V, and the back scan was ended before the anodic steel dissolution. The scan rate was fixed at 20 mV/s.

Morphology and Composition Characterizations

Zn-Mn coatings were electrodeposited under 15 mA/cm^2. After deposition, samples were rinsed with distilled water. Coatings were accurately weighted before and after deposition. Then, the deposits were stripped from the working electrode surface in dilute H_2SO_4 (1 M) solution containing an inhibitor agent hexamethylenetetramine (HMTA) to avoid steel dissolution. The metallic ion Mn^{2+} present in the solution was quantitatively determined by atomic absorption spectrometry (Analytic Jena ZEENIT 700).

The surface morphologies of the deposits were observed using JEOL 5410 Scanning Electron Microscopy. Crystalline structures were analyzed by X-ray diffractometer (XRD) using X-ray D8 Advance Bruker machine with CuK$_\infty$ radiation. The measurements were performed in the 2θ range of 10 to 90°.

RESULTS AND DISCUSSION

Zn Deposition

Cyclic Voltammetry CV studies were carried out to determine the electrochemical behaviors of individual Zn and Mn elements from **S**$_{Zn}$ and **S_{Mn}**. (Fig. **1**) shows the typical voltammograms obtained from S_{Zn} containing only ZnCl$_2$ salt with and without the additive. The peak potentials and peak currents corresponding to Zn^{2+} ions reduction are well identified.

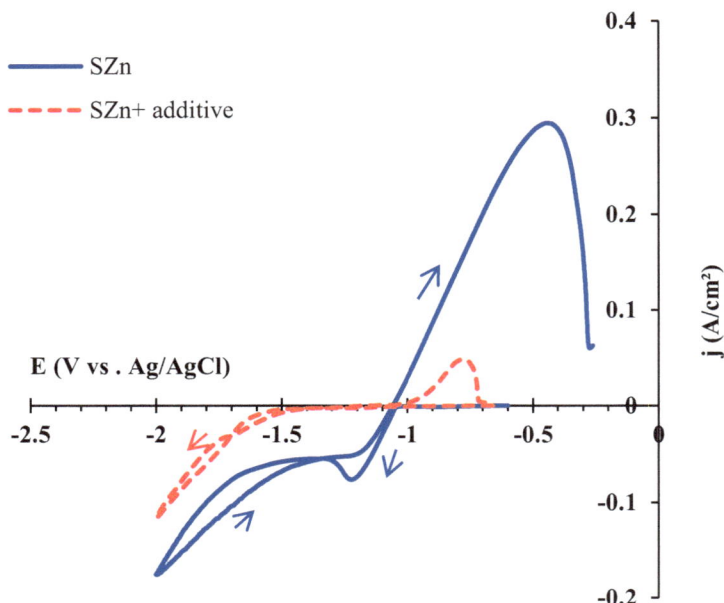

Fig. (1). Cyclic voltammograms obtained from S_{Zn} in the presence and absence of the additive.

In the absence of additive, the cathodic current density j_c sharply increases at -1.03 V *vs.* Ag/AgCl and gives rise to a cathodic peak at around -1.2 V *vs.* Ag/AgCl. This peak is related to the reduction of Zn^{2+} during the cathodic scan. The current density then stays steady between -1.3 and -1.5 V *vs.* Ag/AgCl at about 42 mA/cm². Zn^{2+} ions reduction is controlled by a mass transfer [23]. Above -1.5 V *vs.* Ag/AgCl, a sharp increase of the current density is observed and related to the hydrogen evolution on the Zn film already deposited [23].

Adding the additive (alkylphenol ethoxylated sulphite + BDA) apparently modifies the shape of the voltammogram of Zn. Between -1.03 and -1.55 V *vs.* Ag/AgCl, the cathodic current density j_c is strongly reduced (Fig. **1**). Below -1.55 V *vs.* Ag/AgCl, j_c sharply increases due to the reduction of the metallic ions combined with the hydrogen evolution reaction. The displacement of the cathodic peak to a more negative value is known as cathodic polarization. This decrease in the current density reveals inhibition of Zn^{2+} reduction in the presence of the retained additive. Similar results were reported in previous works [10, 17].

In the reverse scan from -2 V, the anodic peak corresponds to Zn dissolution. The intensity of this peak decreases from 0.3 to 0.05 A/cm². This further result confirms the strong inhibition and the blocking effect of the metallic deposition of Zn during the cathodic scan. According to the literature [24, 25], this blocking effect is ascribed to adsorption of the additive molecules that form an ad-layer on the cathode surface. This ad-layer acts as a barrier blocking most of the active sites, where Zn nucleation occurs. Thus, Zn^{2+} ions discharge is inhibited at this range of potential. An increase in the over-potential is compulsory for desorption of the additive from the cathode surface, allowing Zn^{2+} reduction to occur at the active sites already vacated [26]. This leads to a sharp increase in the cathodic current density at -1.5 V *vs.* Ag/AgCl (Fig. **1**). Moreover, Juhos *et al.* disclosed that BDA is able to increase the capacitance of the double layer [27].

By reversing the sweep towards the anodic direction, two cross-over current densities are observed between the cathodic and the anodic scan. As stated, the appearance of these two cross-over is characteristic of the nucleation mechanism [28, 29]. The cross-over potential value observed in the presence of the additive is more negative than that obtained in the absence of additive (Fig. **1**). This can be due to a lower Zn^{2+} concentration at the interface solution/electrode when the additive is present, as well as the high activation energy for Zn deposition [29].

Mn Deposition

(Fig. **2**) shows the effect of additive on the electrochemical response of Mn on steel from the solution S_{Mn} containing only $MnCl_2$. In the absence of the retained additive, the cathodic current density starts to increase before the equilibrium potential of Mn (-1.57 V *vs.* Ag/AgCl), calculated from the Nernst equation. A wave was observed in the potential range from -0.8 V up to -1.5 V *vs.* Ag/AgCl. According to the literature [30, 31], this wave is related to a hydrogen evolution reaction. At this potential range, Diaz *et al.* demonstrated, using electrochemical quartz crystal microbalance EQCM, that hydrogen evolution reaction alkalizes the vicinity of the working electrode, and an insoluble $Mn(OH)_{2(s)}$ is formed [31]. This is ascribed to the fact that the reduction potential of Mn^{2+} (E° (Mn^{2+}/Mn) = -1.18 V *vs.* SHE) is more negative than that of H^{+} ions.

A slight cathodic peak appears at -1.54 V *vs.* Ag/AgCl. This peak is attributed to Mn^{2+} reduction. Below -1.62 V *vs.* Ag/AgCl, the cathodic current density rapidly increases, showing a broad peak at approximately -1.8 V *vs.* Ag/AgCl. Then, a current hump is probably related to a simultaneous reduction of Mn^{2+} ions and H_2O [30]. Sylla *et al.* also showed that the exact potential at which Mn deposition begins cannot be detected due to hydrogen evolution reduction [19]. Indeed, a review on electrochemical studies dealing with Mn electrodeposition shows a discrepancy due to various solution compositions, purity and/or plating conditions [30 - 32].

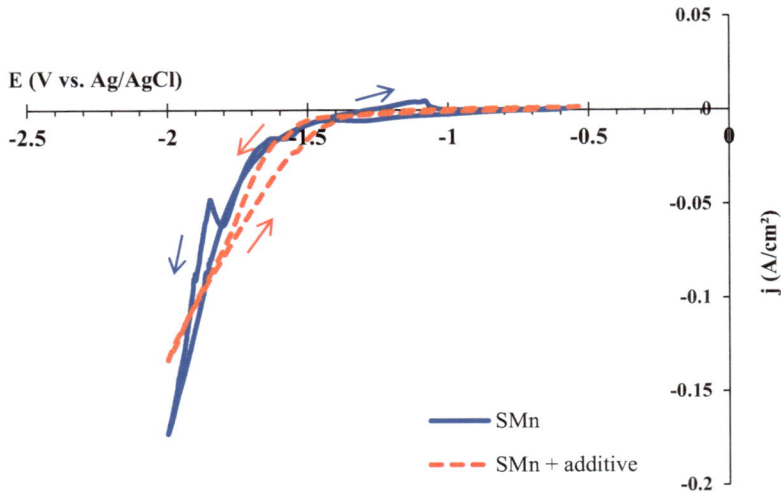

Fig. (2). Cyclic voltammograms obtained from S_{Mn} in the presence and absence of additive.

In the backward scan from -2V, a slight anodic peak corresponding to Mn dissolution is observed at -1.33 V. Such occurrence has been reported by Rudnik for metallic Mn deposited from an acidic chloride bath on a glassy carbon electrode [32]. This is attributed to a spontaneous and rapid chemical dissolution of the metallic Mn in an acidic bath [1, 13, 18, 30, 31]. This non-faradic dissolution simultaneously takes place with the electrochemical one [32] during the anodic sweep. This is due to the low stability of the metallic Mn in acidic media [32].

Introducing the additive (alkylphenol ethoxylated sulphite + BDA) into the solution S_{Mn} affects the voltammogram (Fig. **2**). The cathodic and the anodic features observed in the absence of additive are suppressed; the cathodic peak appeared at -1.54 V is observed. This is due to the inhibition effect of the additive toward the precipitation of $Mn(OH)_{2(s)}$ [31, 32], confirming that hydrogen evolution reaction is hindered.

During the anodic scan, the voltammogram does not display a characteristic anodic peak associated with Mn dissolution. This finding suggests that the Mn phase formed during the cathodic scan in the presence of the tested additive is stable and not accessible for electrochemical oxidation.

Zn-Mn Co-Deposition

(Fig. **3**) compares cyclic voltammetry responses from solution $S_{Zn\text{-}Mn}$ containing both ZnCl$_2$ and MnCl$_2$ salts in the absence and presence of the tested additive.

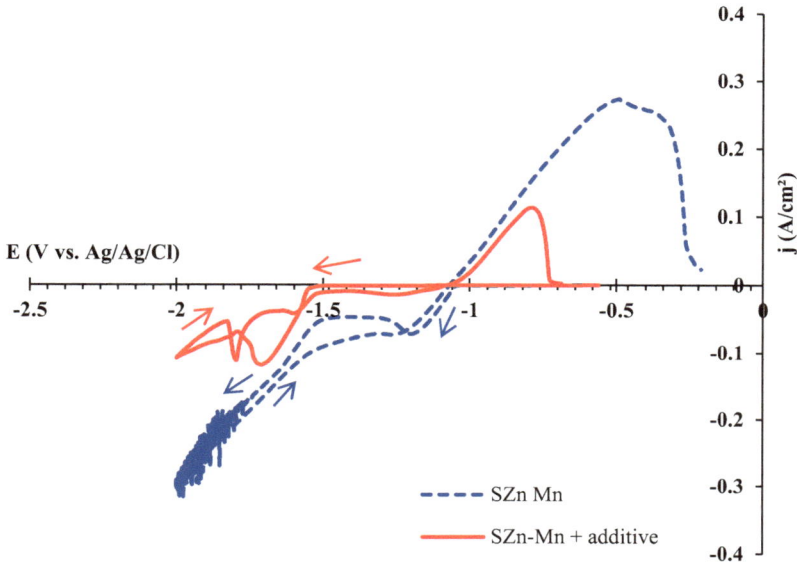

Fig. (3). Cyclic voltammograms obtained from $S_{Zn\text{-}Mn}$ in the absence and presence of the tested additive.

Without an additive, the shape of the cyclic voltammogram is similar to that of pure Zn. The potential at which the cathodic peak -1.03 V (Fig. **3**) arises is close to that obtained during Zn^{2+} reduction when Mn^{2+} ions are omitted (Fig. **1**). This peak is followed by a current hump at approximately -1.52 V *vs.* Ag/AgCl, corresponding to Zn and Mn co-deposition. At more cathodic potentials, important fluctuations of current density are observed from -1.7 V *vs.* Ag/AgCl. This is attributed to both Mn^{2+} reduction and hydrogen evolution via H$_2$ bubbles [19, 33].

When the potential scan is inverted from -2 V to the positive values, an anodic peak with two shoulders is observed. These two shoulders are attributed to the dissolution of different intermetallic phases of alloy [29], evidencing that Zn-Mn alloy has already been obtained during the cathodic scan. It is important to note that the magnitude of the anodic peak (Fig. **3**) is very close to that of pure Zn (Fig. **1**). Moreover, the dissolution potential range of Zn-Mn alloy (-1.08 V to -0.2 V *vs.* Ag/AgCl) is very close to that of pure Zn (-1.08 V to -0.70 V *vs.* Ag/AgCl)

(Fig. **1**). This suggests that the corresponding Zn-Mn alloy contains a high percentage of Zn, proving the preferential deposition of Zn that masks Mn deposition.

The presence of additive in $S_{Zn\text{-}Mn}$ changes the shape of the voltammogram in the following two ways: **(i)** the position of the cathodic peak is clearly different from that seen without additive, and **(ii)** the magnitude of the anodic peak is reduced. The onset of reduction peak is cathodically shifted by almost 500 mV until E = -1.56 V *vs.* Ag/AgCl, revealing an increase in the overvoltage of Zn deposition. This strong inhibition of Zn^{2+} reduction favors Zn and Mn co-deposition.

The potential of the onset of the current density (E = -1.56 V *vs.* Ag/AgCl) (Fig. **3**) is close to that observed for pure Mn deposition (Fig. **2**). All these findings confirm that the onset of the current density is ascribed to both Mn^{2+} and Zn^{2+} reduction (Fig. **3**). Hence, the retained additive (alkylphenol ethoxylated sulphite + BDA) narrows the gap between Zn and Mn deposition potentials.

In the reverse scan from -2 V *vs.* Ag/AgCl, a reduction peak is observed and centered at -1.8 V *vs.* Ag/AgCl, indicating that Mn^{2+} reduction seems to be enhanced on the metallic Mn previously deposited during the cathodic scan. This peak has been pointed out by Sylla *et al.* who studied Mn deposition on Mn substrate [6]. They reported that the Mn deposition peak extends from -1.65 to -1.8 V *vs.* Ag/AgCl during the negative and positive scans. Accordingly, the peak observed during the reverse scan proves a change of substrate to Mn-rich alloy.

In the presence of an additive, a decrease of the anodic peak intensity is observed, revealing that Zn-Mn coatings already deposited contain a lower amount of Zn. This is related to the Zn inhibition effect.

Characterization of the Coatings

Deposits are obtained from $S_{Zn\text{-}Mn}$ in the presence and absence of the retained additive in galvanostatic mode at a steady current density of 15 mA/cm^2.

Chemical Composition of Zn-Mn Deposits

The quantitative chemical analysis results clearly show an increase of the Mn content in Zn-Mn coatings in the presence of the additive (alkylphenol ethoxylated sulphite + BDA). Indeed, $S_{Zn\text{-}Mn}$ provides a pure Zn coating free from metallic Mn, evidencing the fact that it is hard to co-deposit Zn and Mn without additives. Meanwhile, under the same current density, the Mn content jumps to 16.3% with the additive (alkylphenol ethoxylated sulphite + BDA). According to the literature [1, 13, 14, 16], Mn-rich coatings can be obtained only at high current densities ranging from 75 mA/cm^2 [16] to 300 mA/cm^2 [14]. However, the tested additive based on alkylphenol ethoxylate sulphite is able to enhance the Mn content into the Zn matrix, even at a low current density of 15 mA/cm^2.

A review of the literature data shows that such Mn content co-deposited under low current density has not been reported. Similar Mn content of 17.9% was reported in the presence of triton X100 and 3-HBA at a higher current density of 75 mA/cm^2 [17]. The Mn content of the deposits obtained under the same current density as well as the current efficiency of the various working baths are gathered in Table 2.

Table 2. *The Mn content and the current efficiency of coatings deposited from different working baths.*

The Working Bath	Sample Type	% Mn	Current Efficiency (%)
S$_{Zn}$	Pure Zn	-	70
S$_{Zn-Mn}$ (free from additive)	Zn	0	62
S$_{Zn-Mn}$ + (alkylphenol ethoxylated sulphite + BDA)	Zn-Mn	16.3	92.3

The current efficiencies were determined from the chemical analysis of Zn-Mn deposits electroplated from the working baths. According to Faraday's law, the real current efficiency was estimated from the weight gain of the cathode. In the presence of the additive (alkylphenol ethoxylated sulphite + BDA), the cathodic efficiency is about 70% for a high Mn content of 16.3%. This finding attests that the tested additive is able to inhibit hydrogen evolution reaction. According to literature data [14, 16, 20], an enhancement of the Mn content occurs, decrease of the current efficiency. For instance, Crousier *et al.* stated that a high Mn amount of 12% is obtained from the citrate-based electrolyte but with a low current efficiency of 20% [20]. In a similar way, Sylla *et al.* reported that the current efficiency does not exceed 10% if the Mn content reaches 21% [19]. These low current efficiencies are related to the fact that Zn and Mn co-deposition is always accompanied by hydrogen evolution reaction. Compared to literature data, the tested additive has a beneficial effect of favoring Mn^{2+} reduction and inhibiting hydrogen evolution reaction.

Morphological Characterisation of Zn-Mn Deposits

The top surfaces of Zn-Mn alloys electrodeposited from S$_{Zn-Mn}$ with and without the additive were examined. Typical SEM micrographs of the corresponding co-deposits plated by bulk electrolysis are gathered in Fig. (4). Two different magnifications are adopted, 200 μm (Fig. 4 a-c) and 20 μm (Fig. 4 b-d).

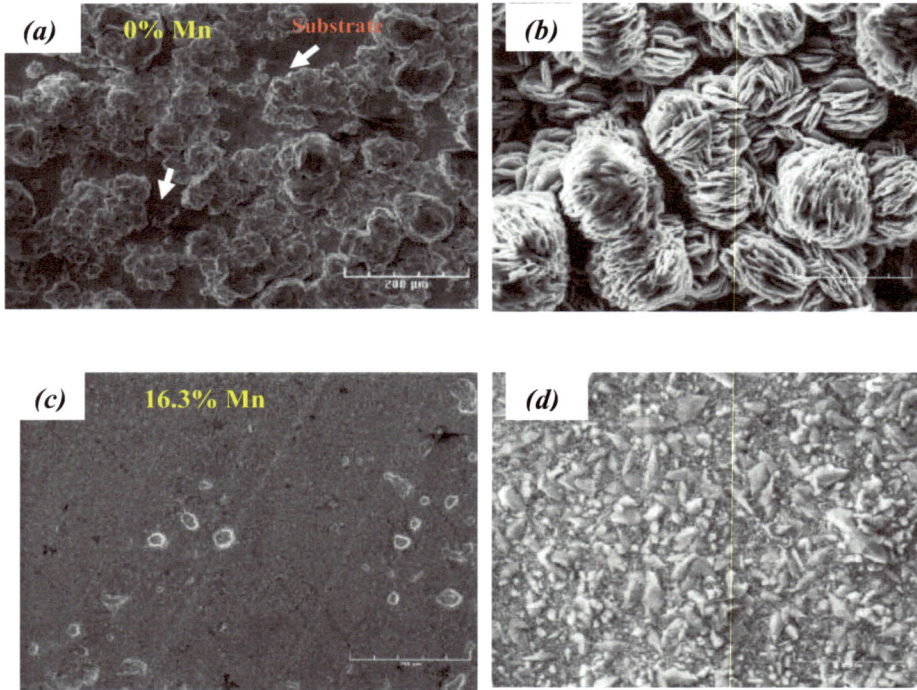

Fig. (4). *SEM micrographs of Zn-Mn alloys obtained from* S_{Zn-Mn}: *(a-b)without additive;(c-d)with additive*

The coating Zn-0% Mn electrodeposited from S_{Zn-Mn} does not cover the whole steel surface, and some parts of the steel substrate can be observed (Fig. **4a**). Diverse grains of various dimensions are irregularly distributed on the substrate surface.

By using high magnification (Fig. **4b**), the crystals of the deposit are shaped as a cauliflower, leading to porous and rough deposits. Similar morphology has been reported for Zn-Mn plated at -1.65 V *vs.* Ag/AgCl from chloride bath [6].

Nevertheless, significant changes in Zn-Mn coating morphology are induced by the additive. Zn-16.3% Mn coating is smooth, continuous, compact, and uniform. However, previous works revealed that Mn-rich alloys containing similar Mn content exhibit dendritic aspect [14, 16]. The Zn-16.3% Mn deposit displays hexagonal pyramid morphology (Fig. **4d**). The morphology of the Zn-Mn deposit varies from flower-like shape (Fig. **4b**) to fine granular shape (Fig. **4d**). This grains refinement also induces compact Zn-Mn deposits. Such morphology is due to the high deposition over-potential brought by the additive, as shown in the electrochemical study (Fig. **3**). This high over-potential promotes nucleation since the free enthalpy of nucleation ΔG decreases, according to the following relation [34]:

$$\Delta G = - nze\,(E - E_{eq}) + \varphi\,(E - E_{eq})$$

(E - Eeq): overvoltage,

φ: surface energy,

z: number of transferred electron,

E: elementary charge,

n: number of atoms transferred;

Accordingly, the number of newly formed nuclei increases, whereas their radius decreases with the increase of the deposition over-potential [34], promoting fine-grained deposit formation.

Structural Properties of Zn-Mn Coatings

XRD analyses were performed to investigate the crystallographic structure of Zn-Mn coatings. The corresponding XRD patterns are presented in Fig. (**5**).

An accurate analysis of the peak positions and comparison with structural parameters reported in the literature [6, 11, 19] allows the identification of Zn-Mn phases.

A pure Zn deposit obtained from S_{Zn} is retained as a reference for comparison (Fig. **5a**). Different diffraction peaks attributed to pure Zn crystals are observed with different crystallographic orientations.

Regarding the deposit obtained from S_{Zn-Mn} (Fig. **5b**), the corresponding XRD pattern is similar to that of the pure Zn deposit (Fig. **5a**), confirming the chemical analysis result (no Mn is co-deposited). Similar peaks corresponding to pure Zn phases are detected (Fig. **5b**). As can be seen in Fig. (**5a**), the most intense line belongs to Zn (101). However, the diffractogram corresponding to Zn-Mn alloy deposited in the presence of additive is monophasic (Fig. **5c**). Only one major peak is observed at $2\theta° = 40.55$ (Fig. **5c**). This phase corresponds to the hexagonal close-packed ε Zn-Mn phase with crystallographic orientation (002) [6, 11, 19]. When Zn-Mn electrodeposition from chloride bath was investigated, Sylla *et al.* demonstrated that ε-Zn-Mn phase electrodeposited on steel appears at $2\theta° = 40.55$ [19].

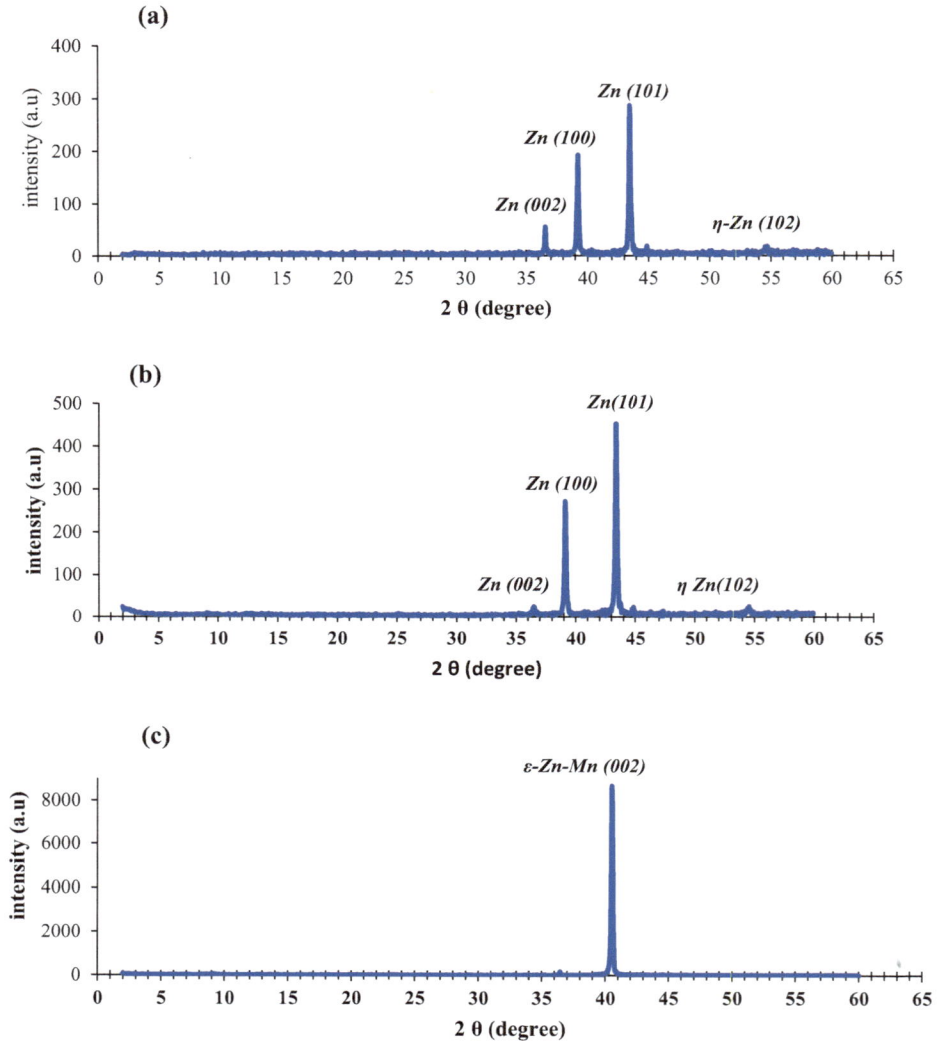

Fig. (5). X-ray diffraction (XRD) patterns of coating deposited at 15 mA/cm^2. **(a)**: pure Zn from S_{Zn}; **(b)**: from additive-free bath S_{Zn-Mn}; **(c)**: Zn-16.3% Mn alloys in the presence of additive.

The pyramidal morphology observed in SEM micrographs (Fig. **5c**) is essentially due to the ε-Zn-Mn phase [35]. It is well documented that the formation of the ε-Zn-Mn phase is beneficial from a corrosion viewpoint [11].

As can be seen in Fig. (**5c**), the most intense diffraction peak (200) suggests that the ε-Zn-Mn crystals are preferentially orientated. By contrast, no preferential crystallographic orientation along a specific direction is observed for Zn-Mn coatings from the additive-free bath [16]. These changes in the crystal orientation are due to a decrease of the metal's surface energy resulting from the adsorption of the additive molecules at the cathode [36]. According to Li *et al.*, selective growth occurs according to the grains which have the lowest surface energy [37].

CONCLUSION

Zn-Mn alloys coatings were successfully electrodeposited on steel from an acidic chloride bath. The electrochemical behavior of each alloying element (Mn and Zn) present individually in the working bath and the mixture of Zn and Mn were investigated in the absence and presence of the additive. The tested additive is composed of alkylphenol ethoxylated sulphite and Benzenlideneacetone BDA. With an additive, the electrochemical data reveals that Zn deposition is significantly shifted until a potential closer to that of Mn. This cathodic polarization attests the inhibition of Zn^{2+} reduction due to the adsorption. Hence, the gap between Zn and Mn deposition potentials is reduced, making Zn and Mn co-deposition possible. The Mn content in Zn-Mn coating deposited at low current density (15 mA/cm^2) reaches 16.3%.

SEM micrographs disclosed that Zn-16.3% Mn coating displays hexagonal pyramid morphology. XRD studies showed that Zn-16.3% Mn coating is monophasic and consisted mainly of the hexagonal close-packed ε-Zn-Mn phase. A preferential orientation is adopted in the presence of the additive. The additive improves the visual appearance of Zn-Mn alloy coatings which are compact and bright. This effect induces a decrease in the number of active sites.

CONSENT FOR PUBLICATION

Not applicable.

CONFLICT OF INTEREST

The authors declare no conflict of interest, financial or otherwise.

ACKNOWLEDGEMENTS

Declared none.

REFERENCES

[1]   C. Savall, C. Rebere, D. Sylla, M. Gadouleau, Ph. Refait, and J. Creus, "Morphological and structural characterisation of electrodeposited Zn–Mn alloys from acidic chloride bath", *Mater. Sci. Eng. A,* vol.

430, p. 165, 2006.
[http://dx.doi.org/10.1016/j.msea.2006.05.025]

[2]   GD Wilcox, and DR Gabe, "Electrodeposited Zinc Alloy Coatings", *Corros. Sci.,* vol. 35, p. 1251, 1993.
[http://dx.doi.org/10.1016/0010-938X(93)90345-H]

[3]   M.A. Pech-Canul, R. Ramanauskas, and L. Maldonado, "An electrochemical investigation of passive layers formed on electrodeposited Zn and Zn alloy coatings in alkaline solutions", *Electrochim. Acta,* vol. 42, p. 255, 1997.
[http://dx.doi.org/10.1016/0013-4686(96)00152-1]

[4]   Z.I. Ortiz, P. Díaz-Arista, Y. Meas, R. Ortega-Borges, and G. Trejo, "Characterization of the corrosion products of electrodeposited Zn, Zn–Co and Zn–Mn alloys coatings", *Corros. Sci.,* vol. 51, p. 2703, 2009.
[http://dx.doi.org/10.1016/j.corsci.2009.07.002]

[5]   M. Mouanga, L. Ricq, and P. Berçot, "Effects of thiourea and urea on zinc–cobalt electrodeposition under continuous current", *J. Appl. Electrochem.,* vol. •••, p. 231, 2008.
[http://dx.doi.org/10.1007/s10800-007-9430-1]

[6]   D. Sylla, C. Savall, M. Gadouleau, C. Rebere, J. Creus, and Ph. Refáit, "Electrodeposition of Zn–Mn alloys on steel using an alkaline pyrophosphate based electrolytic bath", *Surf. Coat. Tech.,* vol. 200, p. 2137, 2005.
[http://dx.doi.org/10.1016/j.surfcoat.2004.11.020]

[7]   B. Bozzini, "Electrodeposition and plastic behavior of low manganese Zn-Mn alloys coating for automotives applications automotive", *Metal finishing,* 1999.

[8]   N. Boshkov, "Galvanic Zn-Mn Alloys-Electrodeposition, Phase Composition, Corrosion Behavior and Protective Ability", *Surf. Coat. Tech.,* vol. 172, p. 217, 2003.
[http://dx.doi.org/10.1016/S0257-8972(03)00463-8]

[9]   N. Boshkov, K. Petrov, and S. Vitkova, "Corrosion Products of Zn-Mn Coatings: Part III. Double-protective Action of Manganese", *Met. Finish.,* vol. 100, p. 98, 2002.
[http://dx.doi.org/10.1016/S0026-0576(02)80445-7]

[10]   P. Díaz Arista, Z.I. Ortiz, H. Ruiz, R. Ortega, Y. Meas, and G. Trejo, "Electrodeposition and characterization of Zn–Mn alloy coatings obtained from a chloride-based acidic bath containing ammonium thiocyanate as an additive", *Surf. Coat. Tech.,* vol. 203, p. 1167, 2009.
[http://dx.doi.org/10.1016/j.surfcoat.2008.10.015]

[11]   C. Muller, M. Sarret, and T. Andreu, "Electrodeposition of Zn-Mn Alloys at Low Current Densities", *J. Electrochem. Soc.,* vol. 149, p. 600, 2002.
[http://dx.doi.org/10.1149/1.1512668]

[12]   D.R. Gabe, G.D. Wilcox, A. Jamani, and B.R. Pearson, "Zinc-Manganese Alloy Electrodeposition", *Met. Finish.,* vol. 91, p. 34, 1993.

[13]   S. Ganesan, G. Prabhu, and B.N. Popov, "Electrodeposition and characterization of Zn-Mn coatings for corrosion protection", *Surf. Coat. Tech.,* vol. 238, p. 143, 2014.
[http://dx.doi.org/10.1016/j.surfcoat.2013.10.062]

[14]   M. Bucko, J. Rogan, B. Jokic, M. Mitric, U. Lacnjevac, and J. B. Bajat, "J, Electrodeposition of Zn–Mn alloys at high current densities from chloride electrolyte", *Solid State Electrochem,* vol. 17, p. 1409, 2013.
[http://dx.doi.org/10.1007/s10008-013-2004-8]

[15]   A. Brenner, *Electrodeposition of Alloys, Principle and Practice 2* Academic Press: New York, 1963.

[16]   N Loukil, and M Feki, "Zn–Mn alloy coatings from acidic chloride bath: Effect of deposition conditions on the Zn–Mn electrodeposition-morphological and structural characterization", *Appl. Surf. Sci.,* vol. 410, p. 574, 2017.

[http://dx.doi.org/10.1016/j.apsusc.2017.02.075]

[17]   N. Loukil, and M. Feki, "Synergistic effect of triton X100 and 3-hydroxybenzaldehyde on Zn-Mn electrodeposition from acidic chloride bath", *J. Alloys Compd.,* vol. 719, p. 420, 2017.
       [http://dx.doi.org/10.1016/j.jallcom.2017.05.142]

[18]   Review—Zn–Mn Electrodeposition, "A Literature Review, N. Loukil and M. Feki", *J. Electrochem. Soc.,* vol. 167, p. 022503, 2020.

[19]   D. Sylla, J. Creus, C. Savall, O. Roggy, M. Gadouleau, and Ph. Refait, "Electrodeposition of Zn–Mn alloys on steel from acidic Zn–Mn chloride solutions", *Thin Solid Films,* vol. 424, p. 171, 2003.
       [http://dx.doi.org/10.1016/S0040-6090(02)01048-9]

[20]   J. Crousier, F. Soto, and M. Eyraud, "Zn-Mn alloys to protect steel plates from corrosion", *Mater. Technol.,* vol. 87, p. 47, 1999.
       [http://dx.doi.org/10.1051/mattech/199987030047]

[21]   M. Bucko, and U. Lacnjevac, "The influence of substituted aromatic aldehydes on Zn-Mn alloy electrodeposition", *J. Serb. Chem. Soc,* p. 78, 2013.
       [http://dx.doi.org/10.2298/JSC130118025B]

[22]   A. Michael Warhurst PhD, *An Environmental Assessment of Alkylphenol Ethoxylates and Alkylphenols,* vol. 26, 1995.

[23]   T. Casanova, F. Soto, M. Eyraud, and J. Crousier, "Hydrogen Absorption During Zinc Plating On Steel", *Corros. Sci,* vol. 39, p. 529, 1997.
       [http://dx.doi.org/10.1016/S0010-938X(97)86101-X]

[24]   P. Sérgio da Silvaa, E. P. Sartori Schmitz, A. Spinelli, and J.R. Garciaa, "Electrodeposition of Zn and Zn–Mn alloy coatings from an electrolytic bath prepared by recovery of exhausted zinc–carbon batteries", *J. Power Sources,* vol. 210, p. 116, 2012.
       [http://dx.doi.org/10.1016/j.jpowsour.2012.03.021]

[25]   C. Ballesteros, P. Diaz-Arista, Y. Meas, R. Ortega, and G. Trejo, "Zinc electrodeposition in the presence of polyethylene glycol 20000", *Electrochim. Acta,* vol. 52, p. 3686, 2007.
       [http://dx.doi.org/10.1016/j.electacta.2006.10.042]

[26]   G. Trejo, H. Ruiz, and Y. Ortega Borgeas, "MEAS, influence of polyethoxylated additives on zinc electrodeposition from acidic solutions", *Jouranl of Applied electrochemistry,* vol. 31, p. 685, 2001.

[27]   S. Juhos, "Mathe, E. Gruenwald, C. Varhelyi and G. Sfintu", *Galvnotechnik,* vol. 83, p. 2282, 1992.

[28]   G. Gunawardena, G. Hills, I. Montenegro, and B. Scharifcker, *Journal of Electroanalytical Chemistry,* vol. 138, p. 225, 1982.
       [http://dx.doi.org/10.1016/0022-0728(82)85080-8]

[29]   V.D. Jovic, R.M. Zejnilovic, A.R. Despic, and J.S. Stevanovic, "Electrodeposition Theory And Practice", *J. Appl. Electrochem.,* vol. 18, p. 511, 1998.

[30]   P. Díaz-Arista, and G. Trejo, "Electrodeposition and characterization of manganese coatings obtained from an acidic chloride bath containing ammonium thiocyanate as an additive", *Surf. Coat. Tech.,* vol. 201, p. 3359, 2006.
       [http://dx.doi.org/10.1016/j.surfcoat.2006.07.152]

[31]   P. Diaz Arista, R. Antano Lopez, Y. Measa, R. Ortega, E. Chainet, P. Ozil, and G. Trejo, "EQCM study of the electrodeposition of manganese in the presence of ammonium thiocyanate in chloride-based acidic solutions", *Electrochim. Acta,* vol. 51, p. 4393, 2006.
       [http://dx.doi.org/10.1016/j.electacta.2005.12.019]

[32]   E. Rudnik, "Effect of gluconate ions on electroreduction phenomena during manganese deposition on glassy carbon in acidic chloride and sulfate solutions", *J. Electroanal. Chem. (Lausanne),* vol. 741, p. 20, 2015.
       [http://dx.doi.org/10.1016/j.jelechem.2015.01.019]

[33] S. Boudinar, N. Benbrahim, B. Benfedda, A. Kadri, E. Chainet, and L. Hamadoua, "Electrodeposition of Heterogeneous Mn-Bi Thin Films from a Sulfate-Nitrate Bath: Nucleation Mechanism and Morphology", *J. Electrochem. Soc.,* vol. 161, p. 227, 2014.
[http://dx.doi.org/10.1149/2.046405jes]

[34] R Ramanauskas, "Structural factor in Zn alloy electrodeposit corrosion", *Appl. Surf. Sci.,* vol. 153, p. 53, 1999.
[http://dx.doi.org/10.1016/S0169-4332(99)00334-7]

[35] B. Bozzinia, E. Griskonisb, A. Fanigliulo, and A. Sulcius, "Electrodeposition of Zn–Mn alloys in the presence of thiocarbamide", *Surf. Coat. Tech.,* vol. 154, p. 294, 2002.
[http://dx.doi.org/10.1016/S0257-8972(02)00010-5]

[36] E Budevski, G Staikov, and WJ Lorenz, *Electrochemical Phase Formation and Growth* VCH Weinheim, 1996.

[37] DY Li, and JA Szpunar, "A Monte Carlo simulation approach to the texture formation during electrodeposition—I. The simulation model", *Electrochim. Acta,* vol. 37, p. 42, 1997.
[http://dx.doi.org/10.1016/0013-4686(96)00164-8]

CHAPTER 9

Prediction of Fire and Smoke Propagation under a Range of External Conditions

Miloua Hadj[1,*], **Blidi Djamel**[1], **Soummar Ahmed**[1] and **Bouderne Hamid**[1]

[1] Department of Mechanical Engineering, Laboratory of Structures Mechanics and Solids LMSS, Faculty of Technology, University Djillali Liabes Sidi Bel Abbes 22000 Algeria

Abstract: The purpose of this paper is to empower the scientific and technological community with the knowledge to identify and define key concepts of fire modeling and to develop the ability to apply the CFD (Computer Fluid Dynamics) tools to fire investigation and prevention using basic mathematical models. Combustion, thermal radiation, turbulence, fluid dynamics, and other physical and chemical processes all contribute to the complexity of fire processes. Flame shape, plume behavior, combustion product diffusion, and thermal radiation effects on neighboring objects can all be modeled with Large Eddy Simulation (LES) software. This paper uses many small and large-scale case studies under various boundary conditions to demonstrate the strength of the Fire Dynamics Simulator (FDS), an LES code established by the National Institute of Standards and Technology (NIST).

Keywords: CFD, Fire scenario, LES, Modeling, Overview, Various scales.

INTRODUCTION

Recent major destructive disasters, such as large-scale forest fires, building fires, large-scale car park fires, and tunnel fires, have led to significant property destruction and, in some cases, devastating human deaths. Fire research applies the theories of science and engineering to investigate the origins, consequences, and prevention of fire.

Fire fighting, fire safety, fire protection, and prevention are the main areas of study in fire science. In addition, fire science encompasses fire management, fire behavior, fire investigation, and hazardous materials. The efforts to understand the fire scenario focus primarily on the behavior of fluids. Fluid flow is about as ancient as science itself. Fortunately, it was not until the 18th century that substan-

---

* **Corresponding author Miloua Hadj:** Department of Mechanical Engineering, Laboratory of Structures Mechanics and Solids LMSS, Faculty of Technology, University Djillali Liabes Sidi Bel Abbes 22000 Algeria; Tel: +213773397141; E-mail: miloua_hadj@yahoo.fr

Zied Driss (Ed.)

tial progress was made in mathematically explaining the fluid flow due to the efforts of Bernoulli, Euler, Reynolds, Poisson, Lagrange, and others. Euler equations define the conservation of momentum for the unsteady fluid and the conservation of mass [1]. For the prediction of various fire phenomena, a variety of models are available. These range from basic algebraic correlations that can only be resolved with a calculator to Zone Models or Lumped Parameter Models that describe space as a small number of elements to CFD or Field Models that estimate a space as a large number of discrete volumes. Zone Models are more advanced than Algebraic Correlations in general, and CFD models are more advanced than Zone Models [2].

Field models, or CFD models, provide a method for modeling the fluid flow through a volume using numerical solutions of the Navier-Stokes equations [3]. The use of CFD models to simulate fire growth and compartment temperatures seems to be the most sophisticated fire modeling approach. The National Bureau of Standards (NIST) in the United States was responsible for designing a CFD model of fire-driven fluid flow named Fire Dynamics Simulator or FDS [4]. This project uses FDS for selected fire problems in different case studies to present an overview of fire modeling.

MATERIALS AND METHODS

The fundamental equations describing the fluid flow and heat transfer mechanisms associated with fires are solved using CFD models. Smoke and heat movement are predicted in forestry, buildings, industrial sites, and every other construction. They are becoming more popular in fire safety engineering. Codes developed specifically for fire engineering applications (also fire prevention code or fire safety code):

- Fire Dynamics Simulator (LES / DNS) [5]
- Jasmine code (RANS) [6]
- KOBRA-3D software [7]
- Smartfire simulator (RANS) [8]
- Sofie code (RANS) [9]

The only model in this category that uses the LES model is FDS, which is the code used in this paper [10]. In fire applications, the LES technique is still expected to be particularly new. NIST, in collaboration with VTT Technical Research Center of Finland, industry, and academics, developed a code named fire dynamics simulator, 'FDS', that has been used to simulate fire and smoke spread problems.

Calculation Tool and Computational Domain

FDS is a computational fluid dynamics model of fire-driven fluid flow. The National Institute of Standards and Technology has made this CFD open-source, and publicly accessible FDS is based on two programs: FDS for simulations and Smokeview for visualizing the results. The user can create one or two text files that include all of the input parameters needed by FDS to define the specific scenario of interest. A fast CPU and a great quantity of random-access memory are needed for FDS (RAM).

The grid size in each simulation and computational domain simulated below have been chosen after testing that results are insensitive to a further refinement of the mesh and to be appropriate and belong to smaller geometry.

On all sides of the domain ($X_1 = X_{min}$, $X_2 = X_{max}$, $Y_1 = Y_{min}$, $Y_2 = Y_{max}$, $Z_1 = Z_{min}$, $Z_2 = Z_{max}$) the boundary condition take a coherent cases:

- passive opening: ambient temperature, ambient pressure and zero velocity gradients;
- Imposed ventilation rate;
- Imposed wind speed;
- adiabatic or no-adiabatic wall;
- ..etc.

At $t = 0$ s, the temperature $T = T_a$, and the flow velocity $V = 0$ m.s^{-1}. The initial time step is $5(\delta_x \delta_Y \delta_z)^{\frac{1}{3}}/\sqrt{gH}$. They depend on the smallest mesh cell $(\delta_x \delta_y \delta_z)$ and the characteristic velocity of the flow $\sqrt{gH}$, domain, g is the acceleration of gravity and H is the height of the domain.

MATHEMATICAL MODELS

The key idea of Large Eddy Simulation applied to the governing equations is to filter out the small-scale motions and to resolve the largest turbulent motions.

A full flow field, $\Phi(\vec{x},t)$, is decomposed into a resolved part, $\overline{\Phi}(\vec{x},t)$, and a subgrid-scale (SGS) part, $\Phi'(\vec{x},t)$, *i.e.*, $\Phi(\vec{x},t) = \overline{\Phi}(\vec{x},t) + \Phi'(\vec{x},t)$. The resolvable part, $\overline{\Phi}(\vec{x},t)$, is calculated from the filtering procedure by taking a function $G(\vec{x}-\vec{x}',\Delta)$ as the filter kernel:

$$\overline{\Phi}(\vec{x},t)= \int_{\Omega}(\vec{x}-\vec{x}',\Delta)\Phi(\vec{x},t)d\vec{x}' \tag{1}$$

Where $\Delta=(\delta_X \delta_Y \delta_Z)^{\frac{1}{3}}$ Δ is the filter width related to the local mesh configuration and δ_X δ_Y and δ_Z the sizes of a grid cell. Eddies of a size larger than Δ are computed, smaller eddies 'unresolved' are modeled or filtered out of the governing equations to each term in the conservation equations, shown below:

$$\frac{\partial \rho}{\partial t}+\frac{\partial \rho \overline{u}_j}{\partial x_j}=0 \tag{2}$$

$$\frac{\partial \rho \overline{u}_i}{\partial t}+\frac{\partial \left(\rho \overline{u}_i\overline{u}_j\right)}{\partial x_j}+\frac{\partial \overline{P}}{\partial x_i}-\rho g_i =\nabla.\overline{\tau}_{ij,SGS} \tag{3}$$

the SGS Reynolds stress tensors is $\overline{\tau}_{ij,SGS}=2u_t\overline{S}_{ij}$, were the local large-scale

rate of strain $\overline{S}_{ij}=\frac{1}{2}\left(\frac{\partial \overline{u}_i}{\partial x_j}+\frac{\partial \overline{u}_j}{\partial x_i}\right)$ and the eddy viscosity can be modelled as

$\mu_t =\rho(C_S\Delta)^2|\overline{S}_{ij}|$. Here, $|\overline{S}_{ij}|$ is the magnitude of the large-scale strain rate

tensor, $\overline{S}_{ij}$, and C_S the Smagorinsky constant [11]. The filtered energy equation for incompressible flow can be done by:

$$\frac{\partial \rho\overline{h}}{\partial t}+\frac{\partial \left(\rho\overline{u}_i\overline{h}\right)}{\partial x_j}-\frac{\partial}{\partial x_j}\left(\frac{\mu_t}{\mathrm{Pr}_t}\frac{\partial \overline{h}}{\partial x_j}\right)=\dot{q}_C''' -\nabla.q_r \tag{4}$$

A perfect gas equation:

$$P_0(t) = \sum_{i=0}^{n} \frac{Y_i}{M_i} = \frac{\Re \rho T}{M} \tag{5}$$

with:

- $\Re$ la constant of perfect gas,

- M_i molar mass of species i (CO2 and fresh air)

Where Pr_t is turbulent Prandtl number.

About combustion modeling, the mixture fraction model is based on the assumption that combustion is incomplete with CO production. The combustion model with two chemical reaction steps is:

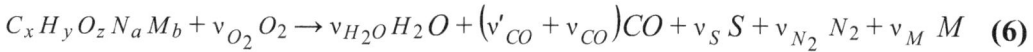

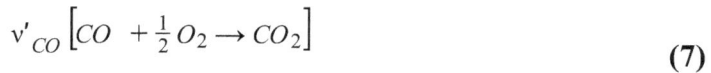

$$C_x H_y O_z N_a M_b + v_{O_2} O_2 \rightarrow v_{H_2O} H_2 O + \left(v'_{CO} + v_{CO}\right) CO + v_S S + v_{N_2} N_2 + v_M M \tag{6}$$

$$v'_{CO} \left[CO + \tfrac{1}{2} O_2 \rightarrow CO_2 \right] \tag{7}$$

The mixture fraction Z to describe the composition of the gas species, have a three components:

$$\begin{aligned}
Z_1 &= \frac{Y_F}{Y'_F} \\
Z_2 &= \frac{W_F}{\left(x - (1 - X_H) v_S\right) W_{CO}} \frac{Y_{CO}}{Y'_F} \\
Z_3 &= \frac{W_F}{\left(x - (1 - X_H) v_S\right) W_{CO_2}} \frac{Y_{CO_2}}{Y'_F}
\end{aligned} \tag{8}$$

Using the mixture fraction via a state relation to derive the mass fractions of reactants and products, $Y_\alpha = (Z_1, Z_2, Z_3)$.

$$Y_F = Z_1 Y_F^l \qquad\qquad Y_{H_2O} = v_{H_2O} \frac{W_{H_2O}}{W_F} Y_F^l (Z_2 + Z_3)$$

$$Y_{N_2} = (1-z) Y_{N_2}^\infty + Y_{N_2}^l Z_1 + v_{N_2} \frac{W_{N_2}}{W_F} Y_F^l (Z_2 + Z_3) \qquad Y_{CO} = \left(v'_{CO} + v_{CO}\right) \frac{W_{CO_2}}{W_F} Y_F^l Z_2 \qquad (9)$$

$$Y_{O_2} = (1-z) Y_{O_2}^\infty - \frac{W_{O_2}}{W_F} Y_F^l \left(v'_{O_2} Z_2 + v_{O_2} Z_3\right) \qquad Y_S = v_S \frac{W_S}{W_F} (Z_2 + Z_3)$$

$$Y_{CO_2} = v_{CO_2} \frac{W_{CO_2}}{W_F} Y_F^l Z_3 \qquad\qquad Y_M = v_M \frac{W_M}{W_F} Y_F^l (Z_2 + Z_3)$$

The stoichiometric coefficients in Eqs.(6-7) are defined:

$$v_{CO_2} = \frac{a}{2} \qquad\qquad v_{CO_2} = x - v_{CO} - (1 - X_H) v_S \qquad v_{CO} = \frac{W_F}{W_{CO}} Y_{CO}$$

$$v'_{O_2} = \frac{v'_{CO} + v_{H_2O} - z}{2} \qquad v_{H_2O} = \frac{y}{2} - X_H v_S \qquad v_S = \frac{W_F}{W_S} Y_S \qquad (10)$$

$$v_{O_2} = v_{CO_2} + \frac{v_{CO} + v_{H_2O} - z}{2} \qquad v'_{CO} = x - v_{CO} - (1 - X_H) v_S \qquad v_S = b$$

Soot is the a mixture of carbon C and hydrogen H, $W_S = X_H W_H + (1 - X_H) W_C$.

RESULTS AND DISCUSSIONS

This research specifically focuses on basic problems in which the heat release rate of a fire is defined a priory and used as input data. It does, however, look into the performance of FDS predictions in more realistic fire simulations where the burning rate is unpredictable and therefore should be computed by the code.

Flame Shape without any External Wind or Ventilation

There are two groups of flames: Premixed flames and diffusion flames. In General, flame generated from a fire can be characterized in terms rate of heat release rate HRR. Conduction, convection and radiation are the three heat transfer mechanisms that can transfer this amount of heat to the surroundings.

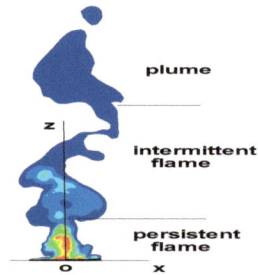

Fig. (1). Flame structure generated from a pool fire without any external conditions.

Also, the natural fires are buoyancy-driven rather than momentum-driven (as being jet flames). A natural fire forms a fire plume which consists of three regions (See Fig. **1**): The persistent flame, the intermittent flame and the buoyant plume [12].

Fire tunnel

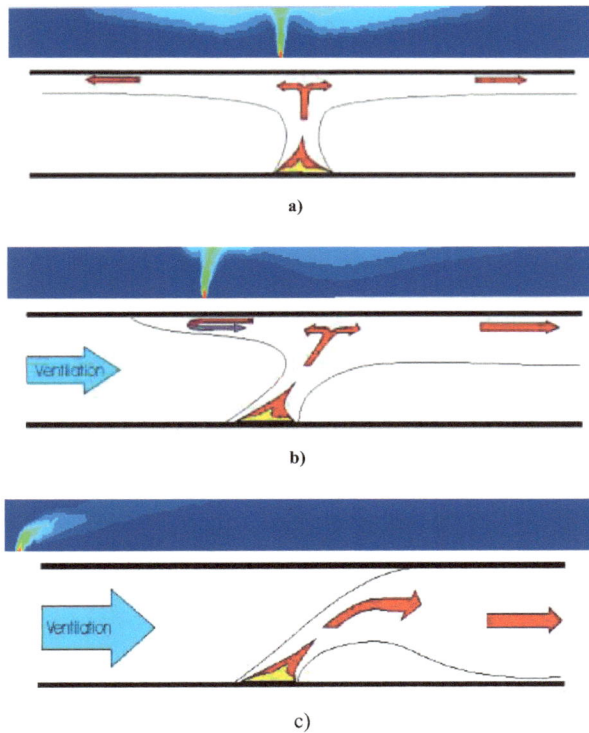

Fig. (2). Impact of the ventilation rate on the form of the flame and the smoke dispersion for fire area D=1 m^2 along the tunnel: **(a)** at V= 0.5m.s^{-1}, **(b)** at V= 0.85m s^{-1}, **(c)** at V= 2.0m. s^{-1}.

By CFD code named FDS developed by NIST, we can understand, predict and model the fire behavior in confined and ventilated road tunnels or similar buildings by characterizing the installations of ventilation and their effect on the flame shape and smoke dispersion (Fig. **2**). Smoke spreads downstream and upstream a fire zone near the ceiling. The present results, although preliminary, indicate that the ventilation rate plays a dominant role in the flame slope and reduces the length of back-layering smoke in the direction opposite to the ventilation L_{RE} (see Figs. **2** and **3**).

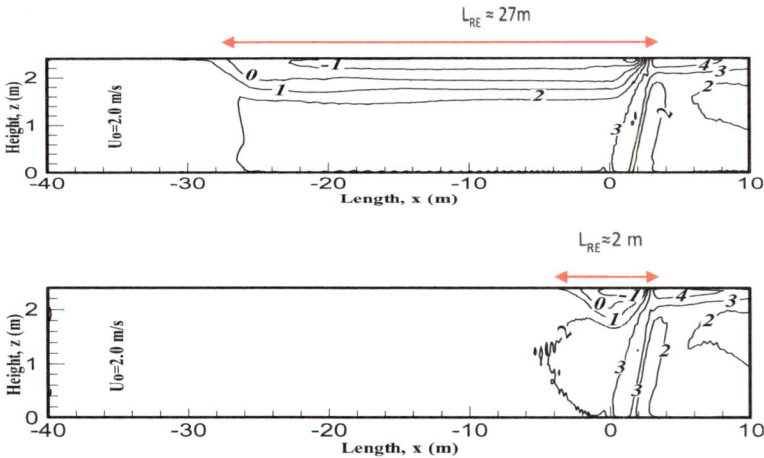

Fig. (3). The effect of thermal boundary conditions in fire and smoke dispersion model generated with different back layer length and ventilation flow rates (adiabatic wall and thermal heat exchange inside wall, respectively).

The value of a critical velocity to prevent the back-layering flow upstream of the fire zone depends on the wall characteristics. Furthermore, L_{RE} is particularly affected by the thermal boundary condition (adiabatic and thermal heat exchange inside wall). Fig. (**3**) shows that a back-layer is suppressed non-adiabatic wall of the tunnel.

Fire and Water Mist Interaction

In several buildings, such as hotels, warehouses, shopping malls and road tunnels, automated sprinkler systems (see left (Fig. **4**)) have to be mounted (see right (Fig. **4**)).

In preventing buildings from fire [13, 14], the sprinkler spray systems which can directly control or suppress the fire are very secure. Bullen [15] has been investigating the relationship between the smoke layer and sprinkler spray. Cooper [16] investigated, using a physical model, the behavior of the smoke layer under the sprinkler spray. One of the most dangerous parts of fire is the smoke dispersion. Like the case of confined fire, it can be dispersed in all directions. Ventilation can have a crucial role in preventing smoke dispersion upstream, downstream, or exhaust smoke from the ceiling, which depends on ventilation mode in the building. The addition of sprinkler or nozzle spray can increase the safe operation by moving the smoke downward instead of being dispersed (see right (Fig. **4**)).

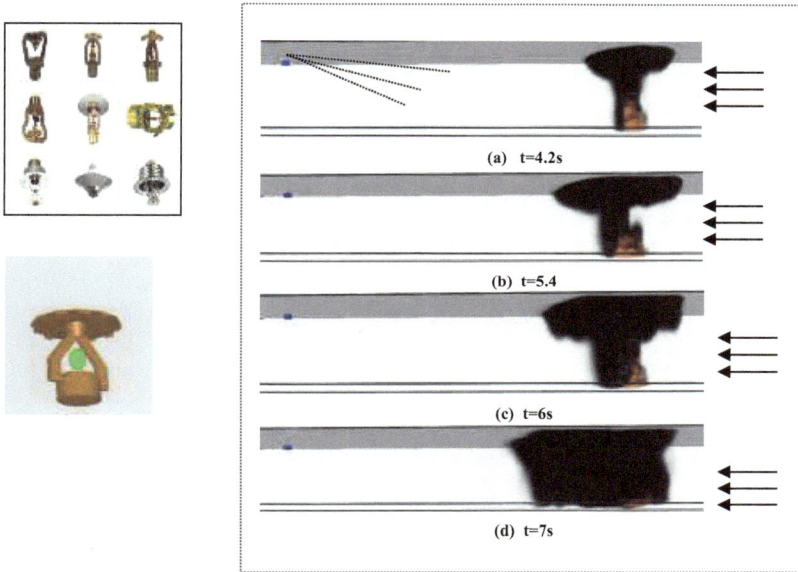

Fig. (4). Plume shape during the interaction with water spray (from sprinkler) downstream fire with HRR of 4MW under critical ventilation.

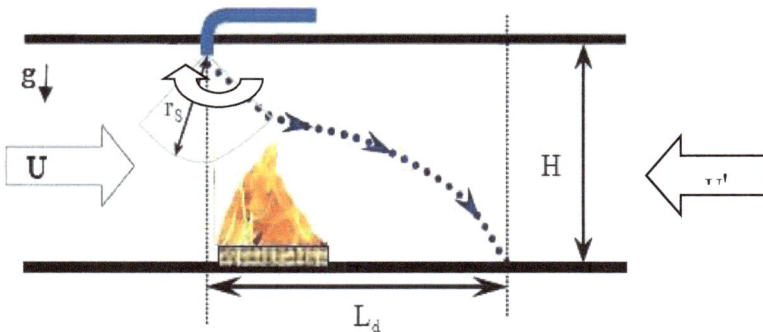

Fig. (5). Droplet water discharged from a sprinkler shall have a landing distance.

Longitudinal ventilation flow can carry the smaller droplets of water spray over long distances (see Fig. **5**). For flows up to 10 m.s^{-1}, the landing distance Ld can reach a value of 50 m recommended by the World Road Association (PIARC). Reducing ventilation speed during the water mist response is an effective alternative for decreasing the L_d [17].

Forest Fire

A rural fire is a wildfire that burns uncontrollably. A wildfire can be categorized more precisely as a forest fire, bushfire, brush fire, desert fire, grass fire, hill fire, peat fire, or vegetation fire, depending on the form of vegetation present. Forest fire (see Fig. **6**) refers to the natural or man-made burning catastrophe that poses a global climate and infrastructure danger, especially to WUI, which are defined as areas where homes are built near or among lands prone to wildland fire [18].

The main factors affecting the fire spread include thermal radiation intensity, wind condition, sloped hill (see Fig. **6**), vegetation characteristics, and weather conditions. Rate of spread (ROS) and heat release rate are parameters used to study fire behavior. On the WUI study, we can also focus on home ignition zone (HIZ). The front fire (see Fig. **6**) can be devised or deformed with the building presence.

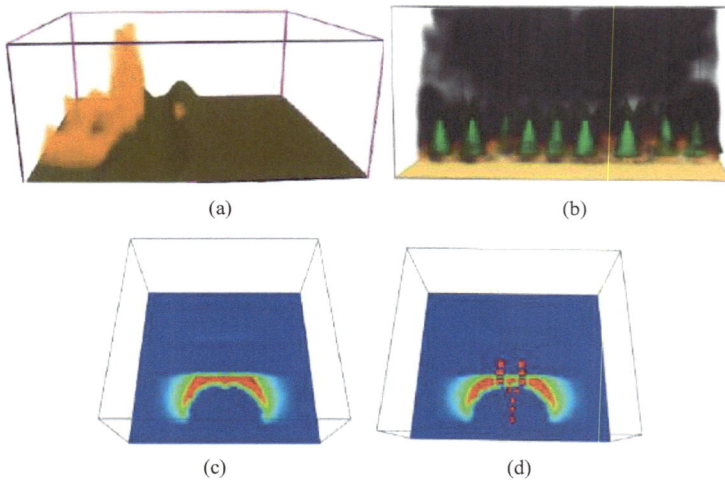

(a)   (b)

(c)   (d)

Fig. (6). Numerical simulation of fire spread on sloped hill and home placed at the top hill and comparison between fire spread with and without the presence of home obstruction.

Domestics and Urban Fire

FDS software used the LES method and performed well in modeling the gas dispersion throughout the building (see Fig. **7**) to determine flammability limit in enclosed areas as well as soot dispersion and the shape of a plume at critical locations and the calculation of fluid flows.

The stack effect, also known as the chimney effect, is caused by air buoyancy and results in the flow of air into or out of buildings (see Fig. **7**). The difference in indoor-to-outdoor air density caused by temperature and moisture variations causes buoyancy.

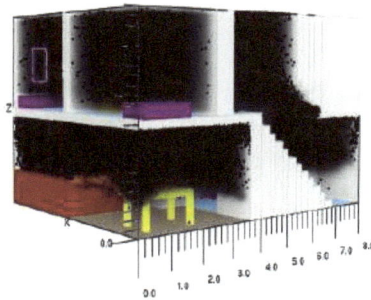

Fig. (7). Fire shape, smoke, and soot particle propagation.

CONCLUSIONS

The implementation of efficient fire safety systems and facilities necessitates an in-depth understanding of combustion phenomena in fires. However, it is also important to apply knowledge gleaned from basic research on the combustion phenomena in fires to the prevention of fires. Combustion is a category of chemical reaction with heat release and heat and mass transfer. Radiation, convection, and conduction are three heat transfer modes responsible for fire spread resulting from burned or unburned fuel.

Smoke flow patterns in buildings are important elements in predicting fire safety. One or all of the following aims can decide the need to maintain the flow of smoke inside a building: Guarantee safe methods of evacuation for the occupants of the building, facilitate emergency services operations, and protect the property.

CONSENT FOR PUBLICATION

Not applicable.

CONFLICT OF INTEREST

The author declares no conflict of interest, financial or otherwise.

ACKNOWLEDGEMENTS

Declared none.

NOMENCLATURE

C_s	Smagorinsky constant
D	diameter of the circular pool fire
$\delta_X, \delta_Y, \delta_Z$	the dimensions of the smallest mesh cell
g_i	acceleration of gravity in the co-ordinate directions x, y and z
h	enthalpy
L^*	characteristic length scale of the fire
P	pressure
P_0	ambient pressure
Pr_t	turbulent Prandtl number
q_m	calorific value of the fuel
$\dot{q}_c'''$	heat release per unit volume
q_r	radiant energy flux
$\dot{q}_r''$	radiative flux to a solid surface.
$\dot{q}_c''$	convective flux to a solid surface.
$\dot{Q}$	total heat release rate HRR
R	universal gas constant
S_{ij}	local large-scale rate of strain
S_{CO}	production rate of carbon monoxide

Sc_t	turbulent Schmidt number
t	time
T	temperature
T_0	ambient temperature
$U = (u,v,w)$	velocity vector
U_{cr}	critical ventilation velocity
U_0	initial velocity
W_α	amount of fuel that is converted to the species α.
y_α	species α yield
X_H	hydrogen atomic fraction
X,Y,Z	cartesian coordinate
Y_{O_2}	ambient mass fraction of oxygen
Y_F^l	fuel mass fraction at the burner surface
$Y_{O_2}^\infty$	fuel mass fraction far from fire zone
$Y_\alpha = (Z_1, Z_2, Z_3)$	mass fraction of species α
Y_F	mass fraction of fuel in fuel stream
w^l	weighting factor in the discrete ordinates method
Z_i	mixture fraction

Greek symbol

u_t	subgrid-scale turbulent viscosity
ρ	air density
ρ_0	ambient air density
$\tau_{ij,\text{SGS}}$	subgrid-scale Reynolds stress tensor
v_α	stoichiometric coefficient of species α

ν_s	stoichiometric oxygen requirement to burn 1 kg soot
ν_F	stoichiometric oxygen requirement to burn 1 kg fuel
Δ	filter width
$\Phi(\vec{x},t)$	general variable
$\overline{\Phi}(\vec{x},t)$	large-scale component
$\Phi'(\vec{x},t)$	subgrid-scale component

Overbar

$-$	filtered variable

Abbreviation

CFD	Computational fluid dynamics
FDS	Fire Dynamics Simulator
HIZ	Home Ignition Zone
HRR	Heat Release rate
LES	Large Eddy Simulation
NIST	National Institute of Standards and Technolgy
RANS	Reynolds Averaged Navier–Stokes
ROS	Rate of spread

REFERENCES

[1]   Garrett Birkhoff, "Numerical fluid dynamics", *Siam Review 25.1 (1983): Waldeck, Benjamin. A Comparison Between FDS and the Multi-Zone Fire Model Regarding Gas Temperature and Visibility in Enclosure Fires,* LUTVDG/TVBB, pp. 1-34, 2020.

[2]   G.P. Forney, and K.B. McGrattan, *User's Guide for Smokeview Version4: A Tool for Visualizing Fire Dynamics Simulation Data.* US Department of Commerce, National Institute of Standards and Technology, 2004.
[http://dx.doi.org/10.6028/NIST.SP.1017]

[3]   M.J. Hurley, D.T. Gottuk, H.K. Hall Jr, E.D. Kuligowski, M. Puchovsky, C.J. Wieczorek, Ed., *SFPE handbook of fire protection engineering.*, 2015.

[4]   J. Ji, L. Zhu, and L. Ding, "Numerical Investigation of External Wind Effect on Smoke Characteristics

in a Stairwell", *Fire Technol.,* 2020.
[http://dx.doi.org/10.1007/s10694-020-00948-4]

[5]   G. Cox, and S. Kumar, "Field Modelling of Fire in Forced Ventilated Enclosures", *Combust. Sci. Technol.,* vol. 52, p. 7, 1986.
[http://dx.doi.org/10.1080/00102208708952565]

[6]   R.B. Basnet, and S. Mukkamala, *Event Detection and Localization Using Sensor Networks.* ICWN, 2009.

[7]   Y-J. Jang, "Comparative Study on The Numerical Simulation for The Back-Layer of The Tunnel Fire-Driven Flow with LES and RANS", *Transactions of the Korean Society of Mechanical Engineers B,* vol. 33, no. 3, pp. 156-163, 2009.
[http://dx.doi.org/10.3795/KSME-B.2009.33.3.156]

[8]   M. Aksit, P. Mackie, and P.A. Rubini, "Coupled Radiative Heat Transfer and Flame Spread Simulation in a compartment", *Third International Seminar on Fire and Explosion Hazards,* 2000

[9]   Noah L. Ryder, "Consequence modeling using the fire dynamics simulator", *Journal of hazardous materials,* vol. 115.1-3, pp. 149-154, 2004.
[http://dx.doi.org/10.1016/j.jhazmat.2004.06.018]

[10]   C.F. Zhang, R. Huo, and Y.Z. Li, "Stability of smoke layer under sprinkler water spray", *ASME's2005 Summer Heat Transfer Confeence,* SanFrancisco, CA, 2005.
[http://dx.doi.org/10.1115/HT2005-72482]

[11]   H. Miloua, A. Azzi, and H.Y. Wang, "Evaluation of different numerical approaches for a ventilated tunnel fire", *J. Fire Sci.,* vol. 29, no. 5, pp. 403-429, 2011.
[http://dx.doi.org/10.1177/0734904111400976]

[12]   V. Raghavan, A.S. Rangwala, and J.L. Torero, "Laminar flame propagation on a horizontal fuel surface: Verification of classical Emmons solution", *Combust. Theory Model.,* vol. 13, no. 1, pp. 121-141, 2009.
[http://dx.doi.org/10.1080/13647830802483729]

[13]   W.K Chow, and B. Yao, "Numerical modeling for interaction of a water spray with smoke layer, Numer", *Heat Transfer,* vol. 39, pp. 267-283, 2001.

[14]   W.K. Chow, and A.C. Tong, "Experimental studies on sprinkler spray–smoke layer interaction", *J. Appl. Fire Sci.,* vol. 4, pp. 171-184, 1995.
[http://dx.doi.org/10.2190/54B4-5AUL-MNCV-F825]

[15]   M.L. Bullen, "The effect of a sprinkler on the stability of a smoke layer beneath a ceiling", *Fire Research Note 1016, Fire Research Station,* Borehamwood, UK, pp. 1-11, 1974.

[16]   L.Y. Cooper, "The interaction of an isolated sprinkler spray and a two-layer compartment fire environment. Phenomena and model simulations", *Fire Saf. J.,* vol. 25, pp. 89-107, 1995.
[http://dx.doi.org/10.1016/0379-7112(95)00037-2]

[17]   R Crosfield, A Cavallo, F Colella, R Carvel, JL Torero, and G Rein, *Travelling Distance of Droplets from Water Mist Suppression Systems in Tunnels with Longitudinal Ventilation Advanced Research Workshop on Fire Protection and Life Safety in Buildings and Transportation Systems,* Santander, Oct, 2009.

[18]   H. Miloua, "Fire behavior characteristics in a pine needle fuel bed in northwest Africa", *J. For. Res.,* vol. 30, pp. 959-967, 2019.
[http://dx.doi.org/10.1007/s11676-018-0676-8]

CHAPTER 10

Structural Design of a 10 kW H-Darrieus Wind Turbine

Soumia Benbouta[1,*], **Fateh Ferroudji**[2] and **Toufik Ouattas**[1]

[1] *Laboratory of Mechanics of Structures and Materials, Department of Mechanical Engineering, Faculty of Technology, University of Batna 2, Algeria*

[2] *Research Unit in Renewable Energy in Saharan Medium, Road of Reggane –Adrar, Algeria*

Abstract: Wind energy is renewable energy that does not require any fuel, does not create greenhouse gases, and does not produce toxic or radioactive waste. Wind power offers the possibility of reducing the operating costs of the electricity system. Vertical axis wind turbines (VAWT) of the Darrieus type, especially in small installations, are increasingly appreciated in current research on wind energy. H-shaped turbines may provide appealing spaces for new design strategies that seek to reduce the visual effect of the rotors and then boost their degree of integration in a variety of installation contexts. The main purpose of this work is to define and critically evaluate the main design parameters of a 10 kW H-Darrieus vertical axis wind turbine that can be considered as a candidate for rural and off-grid urban applications.

Keywords: Blades, Generator, H-Darrieus, Mast, Rotors, Wind turbine.

INTRODUCTION

Wind energy is the fastest-growing alternative energy source in the world, since its purely economic potential is complemented by its large positive impact on the environment. Wind turbines are classified by the orientation of their axis of rotation relative to the wind direction. Two distinct categories can be distinguished in this way: wind turbines with a horizontal axis (Horizontal Axis Wind Turbine, HAWT) and wind turbines with a vertical axis (Vertical Axis Wind Turbine, VAWT) [1].

---

* **Corresponding author Soumia Benbouta:** Laboratory of Mechanics of Structures and Materials, Department of Mechanical Engineering, Faculty of Technology, University of Batna 2, Algeria; E-mail: sousoubenb@yahoo.com

Zied Driss (Ed.)

The first structures designed to generate electricity were the vertical axis wind turbines, paradoxically opposed to the conventional horizontal-axis windmill. Unlike wind turbines with a horizontal axis, wind turbines with a vertical axis are provided with a rotor whose axis of rotation is generally perpendicular to the flow of the fluid, and therefore very often vertical. The technology has existed since the beginning of the 20th century and brings together different technologies: Darrieus, Savonius, Cycloturbine, *etc.*

This type of wind turbine has certain aerodynamic advantages compared to the horizontal axis wind turbine. Some advantages are: no need to put yawn mechanism because it can capture omni-directional wind, simpler design, and lower installation cost due to lower tower top mass [2, 3].

Vertical axis wind turbines (<50 kW) are an ideal opportunity for small-scale electricity production from the wind in urban areas and isolated sites. These types of machines are poorly developed due to the low investment in particular in the mechanical and vibratory design of their structures and also the great development of HAWTs. In particular, the Darrieus type is relatively efficient, robust, requires little maintenance and operates with high wind speeds (> 50 m/s). The design of these types of wind turbines is comparatively simple compared to HAWTs. They are characterized by resonance phenomena induced by the interaction between elastic, inertial and damped forces at specific natural frequencies.

Vertical axis wind turbines (VAWTs) have gained more publicity compared to horizontal axis wind turbines (HAWTs), since they can be built in urban and isolated areas. They can be used at any location, independent of the wind direction (without yawing), and are low-noise, simple to manufacture, and easy to install. Vertical axis wind turbine types include Darrieus, Savonius, and combined Darrieus-Savonius rotors. The H-Darrieus rotor has become popular [4 - 7].

This article deals with the structural design of a 10 kW H-Darrieus vertical axis wind turbine. This work is very interesting for the agricultural development of the Saharan region, isolated sites for various applications, commercial applications such as public telecommunications and radio relay and the contribution to boosting the national economy by the creation of jobs as part of the businesses of young charmers and especially with the economic situation that lives in Algeria today.

The components of the wind turbine structure are geometrically modeled using the industrial software SolidWorks 3D.

TYPES OF WIND TURBINES AND THEIR ADVANTAGES AND DISADVANTAGES [8, 9]

Horizontal Axis Wind Turbines (Hawt) [8]

Advantages:

• High power output

• High efficiency

• High reliability

• High operational wind speed

Disadvantages:

• Difficult to Transport, Install, and Maintain: Due to the sheer size of horizontal axis wind turbines, transporting and installing them come with great logistic and technical challenges.

• Massive tower construction is required to support the heavy blades, gearbox, and generator.

• Stronger tower construction is required to support the heavy blades, gearbox, and generator.

• Reflections from tall HAWTs may affect side lobes of radar installations creating signal clutter, although filtering can suppress it.

• Cyclic Stresses & Vibration – When the turbine turns to face the wind, the rotating blades act like a gyroscope.

Vertical Axis Wind Turbines (Vawt) [9]

Advantages:

• Cheaper to produce than horizontal axis turbines.

• More easily installed compared to other wind turbine types.

• Transportable from one location to another.

• Equipped with low-speed blades, lessening the risk to people and birds.

• Function in extreme weather, with variable winds and even mountain conditions.

• Permissible where taller structures are prohibited.

• Quieter to operate, so they do not disturb people in residential neighborhoods.

Disadvantages:

• Less rotation efficiency.

• Lower available wind speed.

• Component wear-down.

• Less efficiency.

• Self starting mechanism.

WIND TURBINE COMPONENTS

The system studied is H-Darrieus type vertical axis wind turbine of 10-kW power (Fig. **1**). This wind turbine is an electromechanical conversion of wind energy into mechanical or electrical energy based on the principle of aerodynamic differential drag. The wind turbine is designed to convert part of the kinetic energy of the wind into useful mechanical energy gain using a three-bladed wind rotor. The mechanical energy recovered is transformed into electrical energy by a MADA generator (Asynchronous Dual Power Machine).

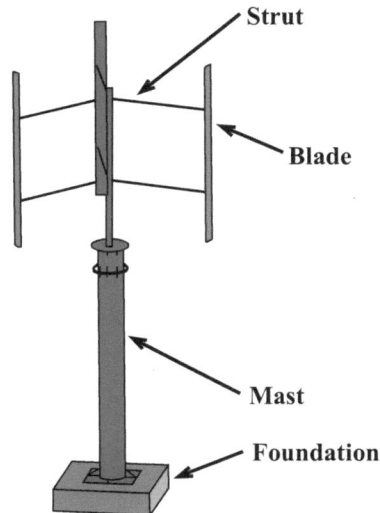

Fig. (1). The Main Components of a Typical H-Darrieus wind turbine.

The wind turbine is made up of three main parts, a fixed part constituting the hollow circular mast, a second intermediate part the MADA generator and a third rotating part composed by the H-Darrieus type rotor. This rotor comprises three blades fixed to the three supports which fixed to the central mast of rotation. The structure of the wind turbine based on a foundation or a concrete base reinforced by anchor rods.

WIND TURBINE DESIGN

The proper set of rotor configurations to be analyzed was then defined. Due to the large number of variables involved in the aerodynamic design of Darrieus rotors [2], some preliminary assumptions were needed to focus the analysis on a significant family of turbines. In particular, the following main choices were made:

- The H-Darrieus configuration with straight blades was selected (Fig. **2**), because of its efficiency and lower manufacturing costs compared to troposkian-blade rotors. The H-Darrieus wind turbine is currently the most widely used and studied solution in the design of wind turbines (Darrieus) [2, 10, 11].
- A blades number N= 3 was assumed. This turbine's architecture guarantees a good efficiency and a sufficiently flat torque profile during revolution, without compromising solidity [2].
- At least two supporting struts for each blade were applied (the number was increased by one strut for each blade whenever too high structural stresses were calculated) [10, 11].
- The comparative study of aluminum and glass-epoxy shows that aluminum has been widely used as a VAWT blade material and glass-epoxy has been used only for HAWT blades and that its application on VAWT blades has not yet been used by manufacturers [12].

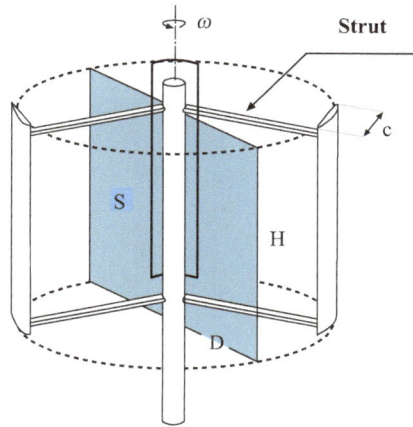

Fig. (2). Schematic view of the architecture of the H-Darrieus turbines [10].

Rotor Design

To design a rotor (H-Darrieus), you must first know the wind turbine operating parameters: the nominal power, wind speed range and the rotor speed. Then, the design parameters: the number of blades (N = 3), Wind turbine swept area (S), the diameter (D) and the height (H) (length of the blade).

Wind Turbine Operation Parameters

a) Specific Speed and Power Coefficient

Current wind turbines tend rather to operate at variable rotational speeds (up to a certain wind speed) so as to maintain a specific speed (λ) constant; it is given by [13, 14]:

$$\lambda = \frac{\omega.R}{v}$$

$$\omega = 2\pi.f$$

(1)

where R is the turbine radius, v is the velocity of incoming air flow, ω is the angular velocity and f is the rotational frequency.

The choice of specific speed for low wind speeds is made taking into account aerodynamic considerations and the noise level. The power coefficient C_p of a wind turbine reaches a maximum value for a given specific speed. In practice, the value of C_p for VAWTs is between 0.2 and 0.4 and λ is between 1 and 5 [15].

b) Design Speed (Nominal)

Wind speed is the most important factor in determining the amount of energy that can be used by a wind turbine. The three wind speed parameters involved in this work are the starting speed (start of production, V_{cut-in}), the nominal wind speed and the speed of production shutdown ($V_{cut-out}$). Jain (2011) expressed that the three wind speed parameters related to power performance as follows [16].

$$V_{cut-in} = 0.5V_{average} \tag{2}$$

$$V_{rated} = 1.5V_{average} \tag{3}$$

$$V_{cut-out} = 3V_{average} \tag{4}$$

All of these speed parameters depend on the average wind speed $V_{average}$.

Power production from a wind turbine is a function of wind speed. The relationship between wind speed and power is defined by a power curve [17]. The Fig. (3) illustrates a typical power curve for a pitch regulated wind turbine. Wind speeds are listed on the horizontal axis, in meters per second (m/s). The turbine's power output is along the vertical axis in kilowatts (kW). In the first region when the wind speed is less than a threshold minimum, known as the cut-in speed, the poweroutput is zero. In the second region between the cut-in andthe rated speed, there is a rapid growth of power produced. Inthe third region, a constant output (rated) is produced untilthe cut-off speed is attained. Beyond this speed (region 4) theturbine is taken out of operation to protect its componentsfrom high winds; hence it produces zero power in this region [18].

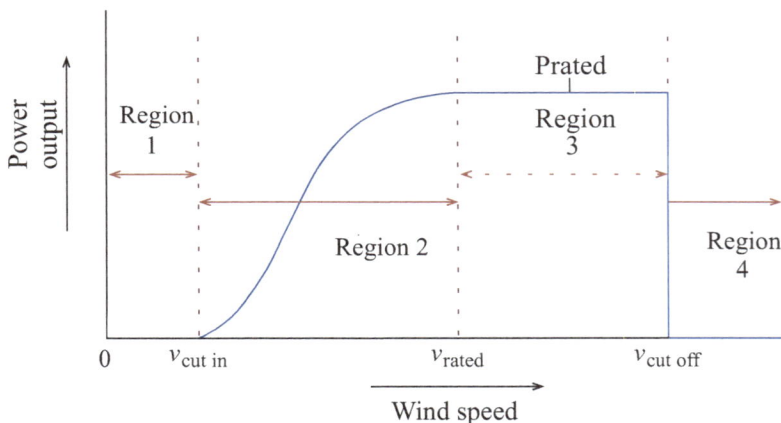

Fig. (3). Wind turbine output power according to different wind speeds [18].

Our wind turbine will be installed in the Adrar region, the average annual speed is 6.3 m/s [19]. Table **1** reported the value of these three wind speed parameters.

Table 1. Speed values; cut-in, rated and cut-out [16, 19].

Parameter	Value
cut-in speed (V_{cut-in})	3 m/s
rated wind speed (V_{rated})	10m/s
cut-out speed ($V_{cut-out}$)	19m/s

Design Parameters

a) The Swept Area (S)

The maximum power that can be developed by a carefully constructed Darrieus rotor, when operating at maximum aerodynamic efficiency is given by the equation:

$$P_{max} = C_p.P_V = \frac{1}{2}C_p(\lambda).\rho.S.V^2 \tag{5}$$

The surface intercepted by the rotor is, then:

$$S = \frac{2P_{max}}{C_p \rho V^3} \tag{6}$$

Where P_{max} in this project is equal to 10 kW,

Practically, C_p is between 0.2 and 0.4 [15]. We will calculate for various values of the specific speed (λ) and the power coefficient (C_p), the diameter and the height of the rotor of our wind turbine [20]. Then, for $C_p = 0.27$, $\lambda = 2.8$ and $V = 10$ m/s the diameter and the height of the rotor are and, respectively. Fig. (**4**) is presented the diameter and the height of the rotor for different values of C_p and λ.

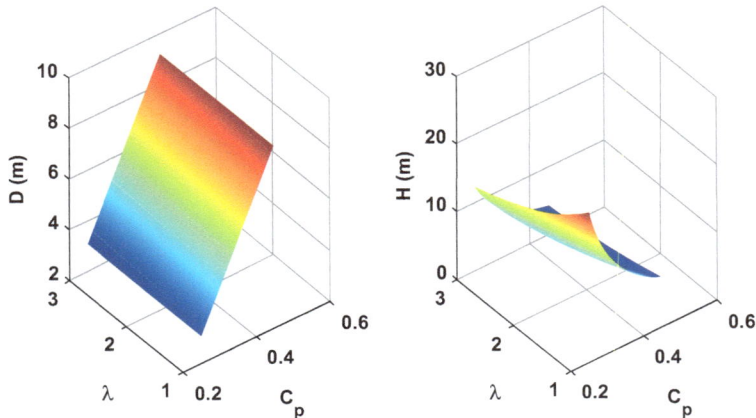

Fig. (4). The Diameter and the height of the rotor (H-Darrieus) of our wind turbine for different values of C_p and λ [20].

The diameter is larger than the height ($D = 1.17H$) to provide a longer chord length (c) for the same Solidity (σ). This design selection provides increased Reynolds number (R_e) for flow on the blades and increased lift.

b) The Solidity

There are several definitions of the structural Solidity (σ). We will define it as the ratio between the total surface of the blades (NcH) and the surface swept by the rotor is expressed by the relation [21]:

$$\sigma = \frac{NcH}{S} = \frac{Nc}{D} \tag{7}$$

c) Blade Profile

Different types of aerodynamic profiles are cataloged around the world. We will note, for example, the Joukowski, Eppler, Wortmann, NACA, RAE, Göttingen, NLR and NASA / LRC and SANDIA profiles. Since the start of its design and development, VAWTs of the Darrieus type have generally been manufactured with blades of aerodynamic symmetrical four-digit profiles of the NACA00XX type, mainly NACA0012, 0015 and 0018, for the high lift, the low drag and good dropout characteristics [12, 22]. NACA profiles are developed by the National Advisory Committee for Aeronautics. This committee existed in the United States from 1915 to 1958 before being replaced by NASA (National Aeronautics and

Space Administration). For our wind turbine, we will take the symmetrical profile NACA 0018. Fig. (**5**) presents the profile whose abscissa and ordinates are dimensioned with respect to the chord of profile (c). Figs. (**6** and **7**) shows the NACA 0018, 2D and 3D profile model in SolidWorks.

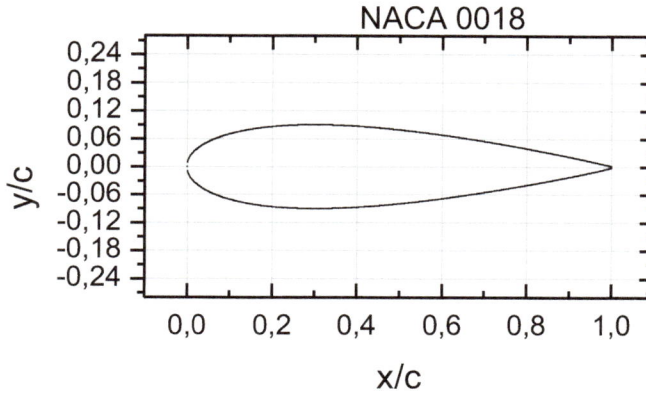

Fig. (5). Profile NACA0018.

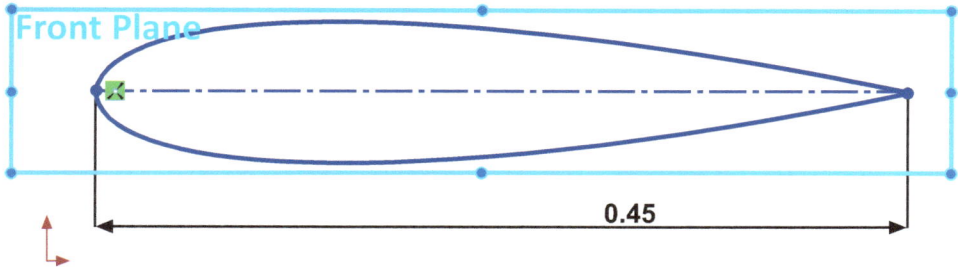

Fig. (6). Profile NACA0018 2D.

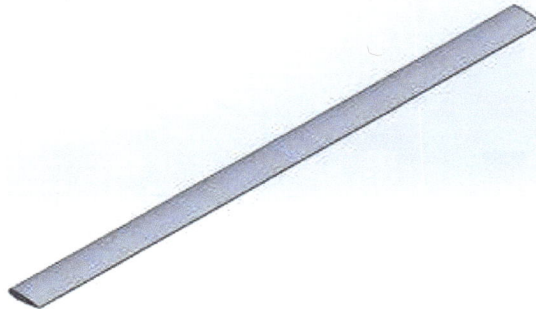

Fig. (7). Profile NACA0018 3D.

Generator Design (MADA)

The electrical configuration of an aerogenerator has a great influence on its operation. The solution with the MADA (Asynchronous Dual Power Supply Machine) whose stator currents are controlled by an inverter and the stator is directly connected to the electrical network, allows the greatest fraction of the power transited by the stator of the machine, which allows reducing the nominal power of the power electronics associated with the machine and therefore reducing the cost of installation.

The fixed speed operation: The electrical system is simpler, reliability is higher, and stability is greater with no need for ordered systems. So the overall system is cheaper.

Through against, variable speed operation allows the increased energy yield, reduction of torque oscillations in the power train, reduces the stress experienced by the power train, and provides better quality of the power produced.

Analysis and diagnosis of the topology of MADA by modeling and simulation: The generator was produced by the project team (Solar and Wind Potentiality Team URER/MS, Adrar) [20, 23 - 25]. The design and modeling geometric detailed at real scale (the sketch of all components, assembly, verification and animation simulation) of the 10 kW generator structure (MADA) is shown in Fig. (8). The geometric parameters of the generator are listed in Table **2** below.

Table 2. Geometric parameters of the generator.

Parameter	Unit	Value
Average radius of the machine	[mm]	554.7
Interior radius of rotor sheets	[mm]	509.3
Outer radius of rotor sheets	[mm]	554.4
Interior radius of stator sheets	[mm]	555.0
Outer radius of stator sheets	[mm]	598.4
No notches (Stator-Rotor)	[°]	1.2
Notches opening (Stator-Rotor)	[°]	0.66
Teeth opening (Stator-Rotor)	[°]	0.54
Spouts opening (Stator-Rotor)	[°]	0.7477

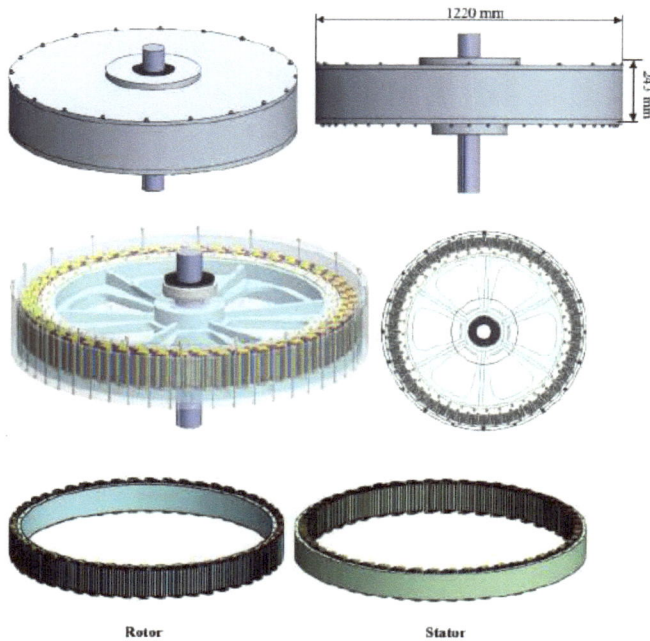

Fig. (8). Generator (MADA) of power 10 kW.

Mast Design

The Current masts are generally made of tubular steel, although it is sometimes found in lattices and some in concrete. In our project, we limit ourselves to the first type. The mast is dimensioned (diameter and thickness) by considering the bending moment created by the thrust of the wind on the rotor and on the mast itself, and by considering the buckling moment due to the weight of the nacelle and the rotor.

The mast is realized using the SolidWorks computer-aided software tool (Fig. **9**).

Table 3. Main technical characteristics of the mast.

Parameter	Unit	Value
Outside diameter	[m]	1
Height	[m]	14
Thickness	[mm]	14

Fig. (9). Mast prototype.

Foundation

The implantation of the wind turbine on the site requires the achievement of a reinforced concrete foundation whose characteristics are recalled below (Table **4**):

Table 4. Technical characteristics of foundation.

Parameter	Unit	Value
Volume weight	[KN.m^2]	25
Width	[m]	4
Length	[m]	4
Thickness	[m]	1

CONCLUDING REMARKS

H-Darrieus wind turbines are gaining popularity in the wind power market, especially since they are considered an appropriate solution even in unconventional installation areas. In this study, an in-depth structural analysis of a 10 kW H-Darrieus vertical axis wind turbine was presented. The components of the wind turbine structure are geometrically modeled using the industrial software SolidWorks 3D. We have identified approximately twenty design parameters in this article, which are: (1) rated power output, (2) rated wind speed, (3) cut-in speed, (4) cut-out speed, (5) the number blades, (6) swept area, (7) clength of the blade, (8) average annual speed, (9) rotor diameter (10) rotor height, (11) material,

(12) blade profile, (13) specific speed, (14) power coefficient, (15) solidity, (16) diameter of the mast, (17) thickness of the mat, (18) height of the mat, (19) geometric parameters of the generator, and (20) technical characteristics of the foundation. It has been demonstrated in this article that these parameters are essential for a 10 kW H-Darrieus vertical axis wind turbine which can be considered as a candidate for off-grid urban and rural application.

CONSENT FOR PUBLICATION

Not Applicable.

CONFLICT OF INTEREST

The author declares no conflict of interest, financial or otherwise.

ACKNOWLEDGEMENTS

Declared none.

LIST OF ABBREVIATIONS

VAWT Vertical Axis Wind Turbine

HAWT Horizontal Axis Wind Turbine

NACA National Advisory Committee for Aeronautics

NASA National Aeronautics and Space Administration

MADA Asynchronous Dual Power Supply Machine

N Number of blades

S Swept area (m2)

D Rotor diameter (m)

H Rotor height (m)

λ Specific speed

ω Angular velocity (rad/s)

F Rotational frequency (s-1, Hz)

V_{rated} Rated wind speed (m/s)

$V_{average}$ Average wind speed (m/s)

V_{cut-in} Cut-in speed (m/s)

$V_{cut-out}$ Cut-out speed (m/s)

P_{max} Maximum power

C_p Power coefficient

σ Solidity

c Chord length of blade

REFERENCES

[1]   X. Jin, "Construction d'une chaîne d'outils numériques pour la conception aérodynamique de pales d'éoliennes, PhD Thesis. University of Bordeaux, 2014", [https://tel.archives-ouvertes.fr/te--01375980]

[2]   I. Paraschivoiu, *Wind turbine design: with emphasis on Darrieus concept.* Published by Polytechnic International Press: Canada, 2002.

[3]   B.A. Dwiyantoro, "The dynamic characteristics of vertical axis wind turbine type H", *J. Eng. Appl. Sci. (Asian Res. Publ. Netw.),* vol. 11, no. 2, pp. 922-925, 2016.

[4]   M. Adaramola, *Wind Turbine Technology: Principles and Design.* CRC Press/Taylor & Francis Group: Boca Raton, FL, 2014.
[http://dx.doi.org/10.1201/b16587]

[5]   F. Ferroudji, C. Khelifi, and F. Meguellatic, "Design and static structural analysis of a 2.5 kW combined Darrieus-Savonius wind turbine", *International Journal of Engineering Research in Africa,* vol. 30, pp. 94-99, 2017.
[http://dx.doi.org/10.4028/www.scientific.net/JERA.30.94]

[6]   R. Kumar, K. Raahemifar, and A.S. Fung, "A critical review of vertical axis wind turbines for urban applications", *Renew. Sustain. Energy Rev.,* vol. 89, pp. 281-291, 2018.
[http://dx.doi.org/10.1016/j.rser.2018.03.033]

[7]   F. Ferroudji, L. Saihi, and K. Roummani, "Numerical simulations on static and dynamic response of full-scale mast structures for H-Darrieus wind turbine", *Wind Eng.,* 2020.
[http://dx.doi.org/10.1177/0309524X20917318]

[8]   "Luvside, Horizontal Axis Wind Turbines (HAWT): Advantages and Disadvantages", 2018.

[9]   "Arcadia/Blog ,Vertical Axis Wind Turbines Advantages & Disadvantages, July 27", 2017.

[10]   A. Bianchini, G. Ferrara, and L. Ferrari, "Design guidelines for H-Darrieus wind turbines: Optimization of the annual energy yield", *Energy Convers. Manage.,* vol. 89, pp. 690-707, 2015.
[http://dx.doi.org/10.1016/j.enconman.2014.10.038]

[11]   A. Bianchini, L. Ferrari, and S. Magnani, "Analysis of the influence of blade design on the performance of an H-Darrieus wind turbine", *Proceedings of the ASMEATI-UIT 2010 conference on thermal and environmental issues in energysystems,* 2010 Sorrento (Italy).

[12]   M. Islam, A. Fartaj, and R. Carriveau, "Analysis of the Design Parameters related to a Fixed-pitch Straight-Bladed Vertical Axis Wind Turbine", *Wind Eng.,* vol. 32, no. 5, pp. 491-507, 2008.
[http://dx.doi.org/10.1260/030952408786411903]

[13]   W. Liu, and Q. Xiao, "Investigation on Darrieus type straight blade vertical axis wind turbine with flexible blade", *Ocean Eng.,* vol. 110, pp. 339-356, 2015.
[http://dx.doi.org/10.1016/j.oceaneng.2015.10.027]

[14]   Y.T. Lee, and H.C. Lim, "Numerical study of the aerodynamic performance of a 500 W Darrieus-type vertical-axis wind turbine", *Renew. Energy,* vol. 83, pp. 407-415, 2015.
[http://dx.doi.org/10.1016/j.renene.2015.04.043]

[15]   B. Liu, and Q.B. Pan, "Large direct-drive wind power system to quickly respond to control strategy", *Electric Mach. Contr. Appl.,* vol. 42, no. 11, pp. 62-66, 2015.

[16]   P. Jain, *Wind Energy Engineering.* McGraw-Hill: New York, 2011.

[17]   "Wind Energy Basics, New York Wind Energy Guide for Local Decision Makers, NYISO: Renewable Resource",

[18]   V. Sohoni, S.C. Gupta, and R.K. Nema, "A Critical review on wind turbine power curve modelling techniques and their applications in wind based Energy Systems", *J. Energy,* vol. 10, pp. 1-18, 2016.
[http://dx.doi.org/10.1155/2016/8519785]

[19]   http://portail.cder.dz, Centre de Recherche en Energies Renouvelables (CDER).

[20]   K. Roummani, M. Hamouda, B. Mazari, M. Bendjebbar, K. Koussa, F. Ferroudji, and A. Necaibia, "A new concept in direct-driven vertical axis wind energy conversion system under real wind speed with robust stator power control", *Renew. Energy,* vol. 143, pp. 478-487, 2019. [http://dx.doi.org/10.1016/j.renene.2019.04.156]

[21]   D.B. Araya, T. Colonius, and J.O. Dabiri, "Transition tobluff-body dynamics in the wake of vertical-axis wind turbines", *J. Fluid Mech.,* vol. 813, pp. 346-381, 2017. [http://dx.doi.org/10.1017/jfm.2016.862]

[22]   Y. Lamine, *Modélisation 3D & analyse du comportement modal d'une pale d'éolienne à axe verticale de type Darrieus, Master memory.* University of Adrar, 2018.

[23]   L. Saihi, F. Ferroudji, B. Berbaoui, Y. Bakou, K. Koussa, F. Meguellati, and K. Roumani, *Hybrid Control Based on Sliding Mode Fuzzy of DFIG Power Associated WECS.,* 2019. [http://dx.doi.org/10.1063/1.5117046]

[24]   K. Roummani, K. Koussa, F. Ferroudji, F. Meguellati, L. Saihi, and Y. Bakou, "A New Study of Direct-Driven Wind Energy Conversion System under Variable Wind Speed", IEEE 2018 6th International Renewable and Sustainable Energy Conference (IRSEC)", *Rabat, Morocco,* pp. 1-6, 2018.

[25]   Y. Bakou, L. Saihi, K. Koussa, F. Ferroudji, Y. Hammaoui, M. Abid, I. Yaichi, and A. Aissaoui, "Design of Robust Control Based on R∞ Approach of DFIG for Wind Energy System, IEEE 2019 1st Global Power", *Energy and Communication Conference (GPECOM), Nevsehir, Turkey,* pp. 337-341, 2019.

SUBJECT INDEX

*9 789814 998208 *